The AMATEUR NATURALIST
and exotic petkeeper

Axolotls
Fruit beetles with attitude
Snakeheads
Solomon Islands Mantids
Striped grass mice
......and much more

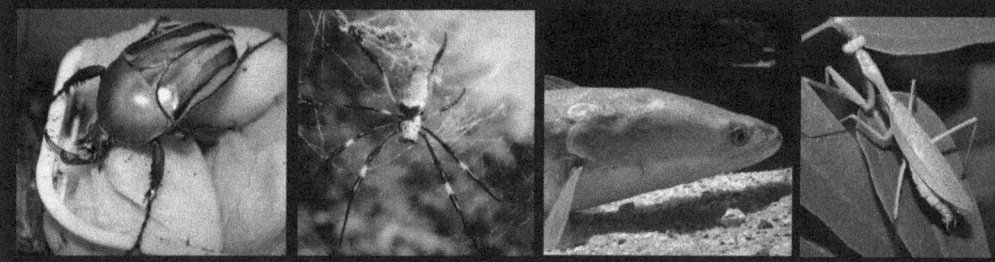

EXPLORE THE WORLD BEFORE IT'S TOO LATE

"In all things of nature there is something of the marvellous."

Aristotle

Edited by Jonathan and Corinna Downes
Typeset by Jonathan Downes, and a certain citrus-coloured felid
Cover and internal layout by Spider for CFZ Communications
Using Microsoft Word 2000, Microsoft , Publisher 2000, Adobe Photoshop.

Published 2009 CFZ Press

CFZ PRESS
Myrtle Cottage
Woolfardisworthy
Bideford
North Devon
EX39 5QR

THE CENTRE FOR FORTEAN ZOOLOGY

www.cfz.org.uk

© CFZ MMIX

ISBN: 978-1-905723-38-6

Plea In Mitigation

Dear friends,

What a peculiar year 2009 is turning out to be. Everything seems to be changing.

We are in the midst of what we are reliably informed is the worst financial crisis in living memory, but at Christmas the shops were full of people spending money like sailors on shore leave.

Frighteningly, a lot of what they were spending money on was livestock, and much of that livestock was exotic.

A few days before Christmas, I was in a Barnstaple pet shop that shall remain nameless, because it's not their fault, and anyway that is where we buy a lot of our animal supplies each month, and I would hate to be banned from there.

However, back to the story.

I was pootling around by the wild bird food trying to find some softbill mix that would tempt the jaded appetites of our pet crested mynah, and jackdaw, when I heard the most appalling dialogue that I have ever heard in a pet shop. I don't usually make a habit of listening to other people's conversations, but I am particularly fond of our colony of African stiped mice, and once I heard somebody else - albeit somebody of the female gender, dolled up to the nines in a revolting Christmas elf costume that was obviously meant to be sexy, but sadly gave the completely opposite impression - mention the species, I completely unethically pricked up my ears. She was haranguing the poor beleaguered shop assistant about the animals she had seen, and more importantly, the animals she wanted to buy her family, and her poor unfortunate boyfriend for Christmas. *"I saw some African striped mice the other day"* she shrieked frenetically. *"They were cute, but I didn't like the colour. Can you get them for me in black and white?"*

I so wanted to pick up one of the long tubes of fat-based wild bird food that was displayed fetchingly on the display stand next to me, and beat her severely around the head and neck, whilst screaming that the colours of wild animals are a matter of thousands of years of evolution, and that the only rea-

Time for a Change

Contents

CONTRIBUTORS

Jon Downes, Graham Smith, Graham Inglis, Corinna Downes, Ray Gabriel, David Loft, Janice Holt, Richard Freeman, David Marshall, Stephen Backshall, Oll Lewis, Ross Phillips, Ray Gabriel, Curtis Lakin, Tim Matthews, Lucy Henson, Trevor Smith, and Max Blake

We make every effort, whilst compiling this magazine, to contact all copyright holders before we use their images. On a few occasions throughout the magazine, this has not been possible, and, if we are contacted by anybody whose copyright we have infringed, we shall pay them an appropriate royalty.

FULL COLOUR DIGITAL EDITION

This magazine is available in two formats. The perfect bound paperback format costing £4.99/$US8.99 and the digital format costing £2/$US4. If you have purchased the hard copy format you are entitled to have a free digital copy.

This is a service of the magazine and its publishers who realise that although the hard copy is more durable and looks well on your bookshelf, the contents are in black and white, so email info@cfz.org.uk with proof of purchase, and download instructions will be sent forthwith.

son one should keep them as pets is to glorify in the magnificence of the natural world. However, it would not have been the act of a gentleman, and I didn't want to get arrested, so I kept my council. It was not until some weeks later that I came to the uncomfortable realisation that if there was enough of a free-market demand for zebra-striped black and white African grass mice then some venture capitalist would probably invest in the genetic engineering mechanism needed to create them.

Sadly, I have seen a lot of things like this in the last year or so. Being the editor of a magazine like this makes you, well it makes me anyway, re-evaluate one's moral stance upon the pet keeping industry. We are no longer involved with the Somerset zoo. I do not want to go into details, but sufficient to say that once again we were forced to re-evaluate our position on a number of issues.

Now, don't get me wrong. I have not become one of the anti-pet keeping lobby, or the anti-zoo lobby, but I believe that it is time for us all, whether private collectors, or zoo keepers to re-evaluate why we keep exotic animals. I have been visiting zoos now for more than four and a half decades. My first word was `zoo` according to my dear departed mother, but Gerald Durrell's mother is also supposed to have claimed that, and I have always suspected that my Mama lifted that eminently quotable line from one of Durrell's books.

The first English zoo that I remember was a long defunct collection in Southampton in 1963, but before that my mother, and the two Chinese women who brought me up used to take me, in my pushchair, to the Hong Kong Botanical Gardens, which was a particularly sordid little collection during the early 1960s, where I would gaze with delight at the rather dishevelled Celebes apes (as they were called in those days), the Chinese water monitor, and a cage of rather bedraggled coatis (one which had a tail missing).

However, I adored the place and have been an avid zoo-goer ever since. I have also kept exotic,

Jonathan Downes
(Editor)
jon@eclipse.co.uk

Graham Smith
(Contributing Editor)
graham.smith19@virgin.net

Corinna Downes
(Sub Editor)
corinna@cfz.org.uk

Richard Freeman
(Herp Editor)
richard@cfz.org.uk

Max Blake
(Associate Editor)
max@cfz.org.uk

Oll Lewis
(Staff Writer)
Oll.lewis@cfz.org.uk

The Amateur Naturalist
Centre for Fortean Zoology [CFZ],
Myrtle Cottage,
Woolfardisworthy,
Bideford,
North Devon
EX39 5QR
Telephone: +44 (0) 1237 431413

Payment by cheque drawn on a UK bank account (payable to CFZ Trust) or by Paypal using our account: jon@eclipse.co.uk.

FOUR ISSUE SUBSCRIPTION
(including p&p)
£18.00 UK
£ 20 (€ 32) Europe
£25 ($US40) US/Canada
£30 ($Aus60) OZ/NZ
£30 Rest of World

or at least unusual, pets since I was about six years old when my first zoological collection consisted of a Chinese praying mantis, and a pair of hummingbird hawkmoth caterpillars in jam jars on my bedroom window. Hong Kong was a great place to grow up when your heart was set on the natural world. However, even from my earliest days I have seen my forays into Nature Study, both in zoos, and in the countryside, and even - let me stress - in jam jars on my windowsill as a spiritual thing; worshipping at the altar of Mother Nature if you will.

And this is where I think we as a community of pet keepers, and zoo goers have gone wrong. For too many people this communion with the natural world has become a mere hobby, or worse a minor lifestyle choice.

Now, we may be committing commercial suicide by changing horses in midstream and transforming what was a relatively successful magazine about exotic pet-keeping into a quarterly journal of radical Natural History, but it was never about the money in the first place. The whole concept of the magazine was to encourage responsible pet-keeping, and to promote the concept of exotic animal husbandry as a part of the study of Natural History, rather than just as a hobby which essentially doesn't go anywhere.

So, in essence nothing has changed. It is still the same magazine that it always has been, except now it is over twenty pages bigger, and we hope better. The title `Exotic Pets` never really sat well with me, and I am glad that we have nailed our colours to the mast with this new re-branding.

We made the decision to change from an A4 magazine format to a perfect bound book one for several reasons. We were printing and manufacturing *Exotic Pets* ourselves, and the task was really too much for us. The overheads were unbearable, and it took such a long time to produce that the whole house got taken over for three nerve-juddering weeks at a time. But the other reason is that these books are more durable; they can be put onto bookshelves like proper books and not left to get crumpled and tatty like magazines, because this a periodical that I think you will want to keep to read again and again, rather than just read once and chuck away.

So welcome to a new, radicalised periodical. We believe that unless we can put our own house in order, then the well-meaning but ultimately misguided folk of the Animal Rights lobby will win, and in twenty years time there will be no exotic pet-keeping (in the UK at least) and every zoo will be exactly the same. No-one will roam the countryside, and children will only see animals as two-dimensional avatars on the screens of their video-games consoles, and that will be a crying shame.

Natural History is no longer seen as a suitable hobby for young people, but in many cases is now an illegal one. A colleague of mine who works for the BBC Natural History Unit told me that children making a documentary on pond dipping were forced to wear safety helmets and rubber gloves before they were allowed near a garden pond. It is now illegal to take frogspawn from your own garden pond and put it in a fishtank.

Changing social mores have meant that most of the children of people I know sit indoors all day playing computer games rather than exploring what little countryside is left. This may seem trivial to you, but Darwin, Linneaus, Mendel, and Gerald Durrell, amongst many others, were amateur naturalists first and foremost. Most professional zoologists started off as amateur naturalists. If kids are no longer able, or encouraged, to do this is it any wonder most of them seem to want to grow up to be image consultants or TV presenters?

We are not condemning the Animal Rights movement. Indeed, during my re-examination of our motives for doing what we do which has taken up much of the last three months, I realised that my attitude towards the people who want to ban exotic pet-keeping, and the people who want to close down zoos has been clouded by the fact that I adore zoos, and have kept what could loosely be described as exotic pets for over forty years.

However, if I am to be honest about it, I have been viewing both lots of pressure groups through my own "filter" of being an exotic pet-keeper who likes to go to zoos. However in the eighteen months or so since I started this magazine (under its previous moniker) I have been travelling around zoos and pet shows, as a journalist rather than as a punter, and whilst I have seen many outstanding and morally uplifting things, other things that I have seen are more than slightly disturbing.

> *"I promise that I will not always write editorials of such enormous length, but the changes in this periodical needed to be explained properly, before we can all go further".*

Too many zoos that I have seen in recent months have been tourist attractions first, and centres of zoological excellence not even second. I am not naïve, nor am I stupid. I realise the commercial constraints under which zoos have to operate, and I also realise that tacky tourist rubbish makes money, but I am becoming more and more uneasy about the philosophical implications of a zoo where the animals are secondary to selling overpriced rubbish to the great unwashed. I can see why the people who wish to have zoos closed down say that such places are philosophically unsound, because by their very nature they are not treating animals with the respect that they are due. And when I see animals being treated as secondary to facepainting, circus skills workshops, and garish, noisy rides, I have to agree.

Too many exotic pet shops, and even shows that I have seen in recent months suffer from a similar malaise. They operate on a sensationalist basis, selling things that look imposing to people who are easily impressed by such matters, whether or not they have any likelihood of being able to successfully look after the poor bloody creatures. I have seen so-called specialist shops where animals are kept under completely inadequate conditions, and sold without even the most basic attempt at giving any help to the prospective buyer. One wonders how many of these poor bloody creatures are going to survive for any length of time. This, too gives plenty of grist to the mill of the people who wish to have the sale of all exotic animals banned.

So, does this mean that I have turned my back on my mindset of the past forty years and have joined the Animal Rights brigade. No, of course not. Just because I understand, and in many cases sympathise with, their grievances, and accept that there is a very real problem, does not mean that I agree with their proposed solutions.

For about a hundred years from the mid 19th Century, Natural History was the most popular hobby for people of all ages, in Britain and her Empire. And as I wrote earlier, the great names of zoology and conservation of the 19th and 20th Century were nearly all, originally at least, amateur naturalists, and some - like Gerald Durrell, for example - never attained conventional scientific qualifications. Darwin, as you will read elsewhere in this issue, in an article to mark the bicentennial of his birth, flunked his medical training, took a degree with the intention of becoming a parson, and then became the greatest zoologist of the last 200 years, without a formal zoological qualification in sight, and even I have no zoological qualifications apart from a not very good `O Level` from 1976.

I can imagine a world without the next generation of Jon Downeses in it, but a world without a 21st Century Darwin, Mendel, Linnaeus, or Durrell. That would be unthinkable.

And that is what I believe will happen if our increasingly urbanised and sedentary population are not encouraged to take an active, rather than a passive, interest in the natural world. And this is exactly what I believe will happen if Exotic Petkeeping, and zoos with anything like an appropriate agenda do not continue in this country. And, unless we as a community regulate and police ourselves this is exactly what is going to happen.

It is time not only for a resurgence in the somewhat neglected occupation of the Amateur Naturalist, but it is time for the emergence of the Radical Naturalist. It is time to take a stand against the manifest stupidities and injustices which beset us at every turn. It is time to insist that those who want to keep exotic animals do so for the right reasons; in order to study their behaviour and habits, and to in their own little way add to the sum total of human knowledge. It is time to insist that zoos, petkeepers, and those who - like us - are part of the petkeeping industry treat the animals with which we deal with the respect and awe that they deserve. It is time that we stop treating the Animal Rights extremists as the unquestioned enemy and try to establish a dialogue with them, because although their methodology and beliefs are different to ours, their ultimate intention - to treat the other denizens of the world in which we live with respect and dignity - is the same. And it is time to realise that if we don't take a stand, and furthermore, take a stand now, then the way of life that we hold dear will not exist for much longer!

I promise that I will not always write editorials of such enormous length, but the changes in this periodical needed to be explained properly, before we can all go further.

Onwards and Upwards,

Jon Downes
Editor, *The Amateur Naturalist*
North Devon,
February 7th 2009

Dr Karl Shuker is perhaps Britain's best loved, and most respected cryptozoological writer. CFZ Press are proud to be publishing a series of his books, mixing new titles with updated reissues of his classic works.....

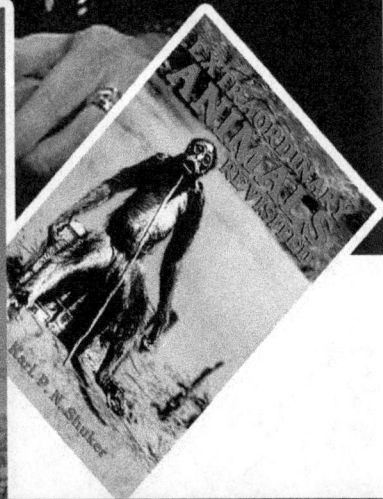

The News

Compiled and Edited by Jon Downes and Oll Lewis

MARSHING MORONS

Y ou'd expect the second best invertebrate conservation site in the United Kingdom to be protected from development, wouldn't you?

Sadly, this is not the case. West Thurrock Marshes in South Essex is due to become a warehouse and lorry park if developers get their way.

The site is home to 1200 different species of insect, bird and reptile, some of which are rare. It is also a stunning example of succession in action; the site has been transformed from a being a power station and ash dump in the 1980s into a wetland and grassland. Grasslands are among the rarest habitats in Britain, and are far different to a garden lawn or a regularly grazed field. Lawns and fields, because they are regularly cut and grazed during the growing seasons, never attain the level of diversity in plant life that a grassland does. Despite its rarity and importance to many

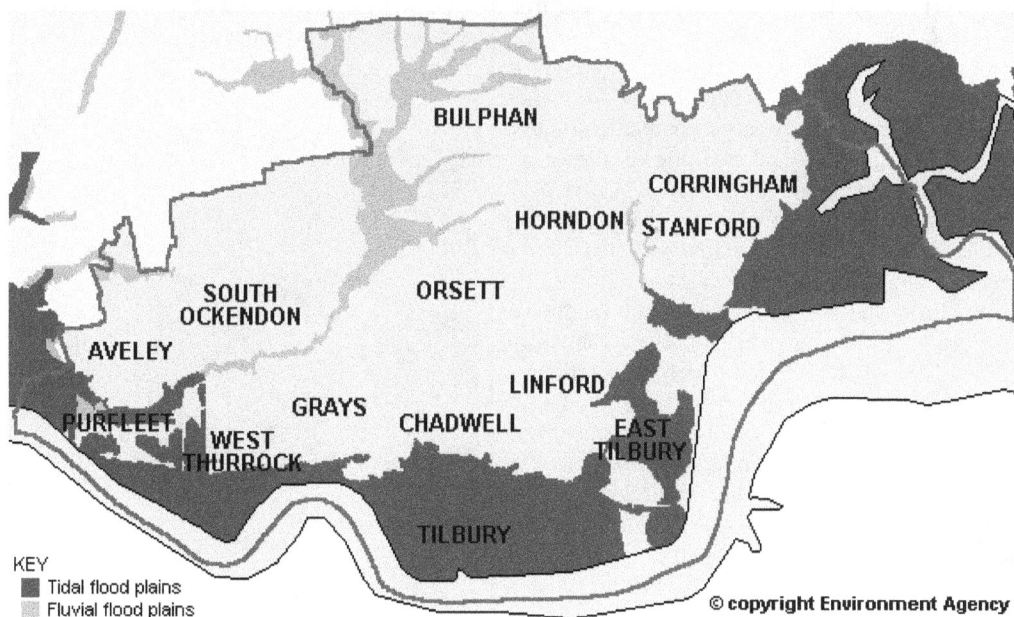

© copyright Environment Agency

Brownfields are abandoned or underused industrial and commercial facilities available for re-use. The more formal term for brownfields is "previously developed land" (PDL), the definition of which talks of it being vacant, derelict or underused. It may not have been industrial in the past, and it may or may not be contaminated.

The government has a target that 60% of new housing development must be on PDL,[citation needed] and the overall aim in the country is to recycle PDL in preference to taking greenfield sites.

rare species, especially to butterflies, less than 1% of all grassland in the United Kingdom is protected by nature reserves and an even smaller percentage is actually owned by any kind of nature conservation trust or charity, which is the only way to guarantee the land is completely safe from development.

Conservation sites are usually leased from their owners or set aside by large companies or land owners because the land, for one reason or another is not commercially viable. This set aside land is often old brownfield sites that are given a warden who is funded by grants provided by successful applications or environmental action plans. From the point of view of the landowner this is a good thing as it helps prevent vandalism, is certainly good PR and rarely costs them a penny if the sites upkeep can be funded by grants. The benefits to wildlife are obvious, especially to species on the local Biodiversity Action Plan (BAP) and the sites are important to biologists who can monitor species and population dynamics on these sites. One site I used to visit on fieldwork when I was in university was adjacent to

Liverpool Dock, and was owned by the port authority. The site had fallen into disuse due to the shrinking demand for the port after Britain's potentially still very profitable coal industry was all but shut down by the Margaret Thatcher's Conservative Government merely to send a message to powerful workers unions. This site is not pleasing to the eye, being a brownfield site, but like West Thurrock Marshes is home to a vast array of wildlife, rare in most other parts of the country.

Old brownfield sites like West Thurrock are not just important because they are grassland but because of the large amounts of stone you'll find there. In old brownfield sites the ground is usually littered with rubble from the buildings that once stood there and this is tremendously good for invertebrates.

Thanks to Greenbelt, though, as much as half of West Thurrock Marshes is to be lost to planned building works. Greenbelt, despite its environmental sounding name is nothing to do with environmental protection, it is just a scheme cooked up by governments to ensure that people in the Home

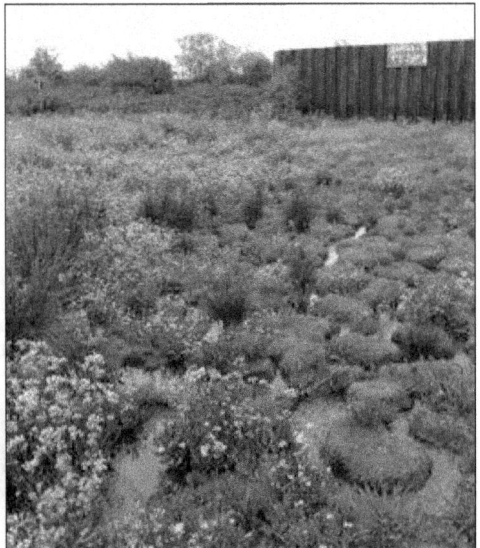

Counties didn't get worried that their little towns would get swallowed up by the urban sprawl of London. Because of this nimby-ism in great swathes of the South East of England it is difficult to build new residential and business properties on land that would be suitable for this purpose and of no special environmental interest. This is were brownfield sites come in: if a site was previously approved for development by a local council then if what is currently on the site is derelict or out of use it is usually very easy to gain planning permission to develop on it. This is what has happened to West Thurrock Marshes. Also, as is the case all too often with planning applications, the development companies have sidestepped environmental concerns

"This is a disappointing decision which reveals the inadequacy of our current wildlife protection. What right do we have to ask other countries to protect their rainforests or coral reefs while we continue to destroy the most valuable habitats of our own endangered species?" says Matt Shardlow, Buglife Director. *"The Government must act now to strengthen its biodiversity legislation and halt the worsening loss of wildlife".*

by setting aside a small part of the site to continue as a nature reserve. This sort of thing usually impresses the legal profession, who usually know or understand very little about ecology and are easily fooled or bamboozled by halfhearted bone throwing like this. The plan does not take into account that you cannot just move the species present in the different habitats of one-half of the site to the other, nor will they automatically move of their own accord. Even if a species of insect were trans-located to another part of the site, its population would drop considerably due to competition with other species in the immediate area.

The charity Buglife recently lost an appeal against the development and will have to pay all legal costs, so it looks like the plan will go ahead. Another worry is that there would be nothing stopping the company that owns the land mounting a similar planning application ten years or so down the line saying they'll save a portion of the remaining half of the grassland. This could happen to any nature reserve, but brownfield sites are particularly vulnerable to this because often the public and the professionals who make the decisions will often just think of the sites as being eyesores, even if the site in reality is quite beautiful like West Thurrock Marshes, and not being potentially very important for conservation.

How to help protect brownfield nature reserves

If you want to ensure that brownfield sites are given greater legal protection the only way you can do this is by lobbying your local Member of Parliament to take this issue seriously, this is how laws are made. Another way to get your voice heard is to join your MP's party and raise the issue of protection of brownfield nature reserves at constituency meetings.

If you can, get everyone you know to write to their local MPs too and get them to ask everyone they know to write to theirs. If we sit back and take this then it is not going to go away, it'll just get worse and worse. The time for action is now and the cost is that of a second class stamp, together we can change the law and make things that little bit better for animals and future generations of people, what are you waiting for?

Remember to always be polite in your dealings with the MP, even if you don't agree with them, or the party they represent. Campaigns are most effective if the MP believes you to be fair and balanced and someone who might possibly vote for them. Include your return address so the MP can reply to you and to prove that the letters he is receiving are not all coming from one person driving around all the post boxes in the area. If you are unsure of what to write feel free to use the following text:

Dear (MP's name),

It has come to my attention that nature reserves situated on brownfield sites enjoy no protection under UK law. Planning permission can be given to tear up important habitats that are home to rare and endangered species when development companies offer half-hearted and inadequate contingency plans for wildlife.

I believe better legal protection should be given to nature reserves situated on brownfield sites, to counter the erroneous belief that these sites are unimportant to wildlife and merely a blight upon an area.

One of the species which thrives in the Thurrock Marshes is *Bombus humilis* - a brown carder bumblebee which makes its nest on the surface of the ground at the base of long vegetation, often under accumulated plant litter. It has most often been recorded as associated with areas of grassland supporting a large number of plant species with long corolla flower types, such as the plant families Lamiaceae and Fabaceae. It is one of a number of bumblebee species to have undergone a drastic reduction in range and abundance, as a result of the loss of this habitat in the modern agricultural landscape, although it appears to be able to survive in less extensive areas of flower-rich habitat compared with some bumblebee species.

CONTACT: Buglife - The Invertebrate Conservation Trust, First Floor, 90 Bridge Street, Peterborough, PE1 1DY. Tel: (0)1733 201210

Our Own Little Mobile Clone

For some years scientists have been working on various conservation projects involving the use of cloning. The most well known of these are projects are attempts to clone species that are believed to be extinct, like the woolly mammoth and the thylacine. However, due to the difficulty inherent in obtaining suitable DNA samples, plans to resurrect these species have failed. In order to resurrect the mammoth it would be necessary to be able to produce clones of animals from tissue that had been frozen without the use of cryopreservation chemicals. Normally this would be impossible because of damage caused by ice-crystals to the DNA and cells of animals. However, a team of Japanese researchers working under Teruhiko Wakayama in the RIKEN research institute in Kobe, Japan appear to have perfected a cloning system that allows them to clone animals from frozen cells. The team cloned mice from brain cells of a mouse that had been frozen for 16 years at -20 degrees C without the use of cryopreservation chemicals. Dr Wakayama said that this improved process would offer a distinct chance to resurrect extinct species, or to preserve endangered species.

The Amami Rabbit also known as the Ryukyu Rabbit, is a primitive dark-furred rabbit which is only found in Amami Ōshima and Toku-no-Shima, two small islands between southern Kyūshū and Okinawa in Kagoshima Prefecture (but actually closer to Okinawa) in Japan.

Often called a living fossil, the Amami Rabbit is a living remnant of ancient rabbits that once lived on the Asian mainland, where they died out, remaining only on the two small islands where they survive today.

The Amami Rabbit has short legs, a somewhat bulky body, rather large and curved claws, and is active at night. Its ears are also significantly smaller than those of other rabbits or hares.

While still on the subjects of cloning and Japan, scientists at the Kinki University in Osaka have recently succeeded in cloning the endangered amami rabbit (*Pentalagus furnessi*) from an animal that had been found dead. It is hoped that the animal will survive considerably longer than some previous examples of cloned endangered animals like the Asian gaur, a rare Indian ox that survived for only two days after it was born in 2001.

Frogs eaten to extinction

David Bickford of the National University of Singapore has warned that the global trade in frog meat could cause some species to go extinct. France and America import large amounts of frog meat from the Far East with France alone accounting for up to 4000 tonnes a year. Frog is also a popular meat in Asia and Bickford estimates that 180 billion frogs are harvested each year from the wild and if the global frog meat trade continues unregulated and unchecked the way it has been many species of frog will soon become extinct.

Rare Chelonian is Turtley Safe

One of the world's rarest turtles was saved from the cooking pot on the 26th of November in Vietnam. The turtle, one of only 4 known living individuals of Swinhoe's soft-shelled turtle (*Rafetus swinhoei*), the most famous example of which being the gigantic turtle of Hoan Kiem Lake in Hanoi, Vietnam which, if local folklore is to be believed, is over 500 years old.

The turtle, whose shell alone is over 1metre long, was netted by fishermen in Son Tay Vietnam. Conservationists, who had been monitoring it in Dong Mo Lake before flooding enabled its escape into a local river two weeks ago, were quick to arrive on the scene and attempted to save the turtle. The animal's captor, however, had other plans and had been offered 30 million Dong (US $4,800) for the animal by a local restaurant owner. He then offered the turtle to the conservationists for $1,400 and this led to a protracted negotiation process involving the fisherman brandishing a stout stick. Eventually a large crowd of over 100 onlookers and tourists arrived to watch the proceedings. Many of them were concerned about the welfare of the turtle, and the fisherman agreed to return the turtle to the conservationists for a $200 finder's reward and new nets.

One of the conservationists, Nguyen Thi Van Anh from Education for nature, said:

"It's a small reward for people who realized that

it's important to protect one of the most endangered species in the world."

Conservation efforts with this species are concentrated on breeding captive turtles in China and searching for live specimens in the wild. An agreement was made to transfer the only known remaining female specimen located at the Changsha Zoo to the Suzhou Zoo to breed with the male specimen there. Also efforts are being made to improve conditions for breeding at both the Suzhou Zoo and Western Temple in Suzhou. A workshop on the Rafetus Conservation at Yunnan was held by CI-Shanshui. Local Chinese scientists are searching for the last existent individuals. The two specimens were able to produce two clutches of eggs with over half of them being fertile, though unfortunately all of them perished before hatching. The Turtle Survival Alliance released a statement, saying *"A number of the eggs had very thin shells, suggesting that the diet of the animals prior to breeding was not optimal."*

The two turtle are now being prepared for another round of mating, while being fed a high calcium diet in an effort to strengthen the eggs. Liu Jinde, the director of the zoo said *"We've worked very hard on this, We ought to succeed. The turtles are very healthy."*

The scientists are preparing to mate the two once again in May 2009, which falls within this species' breeding season.

ZOO NEWS

COPYRIGHT
BVZ 2
ELEPHANT RIDE
Zoological Gardens, Belle Vue, Manchester
LILYWHITE LTD
BRIGHOUSE

Enter the dragonfish

Bristol Zoo has become the first zoo in Britain to breed dragon-fish (*Scleropages formosus*) in captivity. The dragonfish, also known as Asian arowana, have been listed as endangered by the IUCN since 2006 and have been listed under appendix 1 of CITES since 1975.

Due to their rarity, caused mainly by habitat destruction on the Malay Peninsula the fish is highly prized by less scrupulous aquarists and can fetch high prices for live specimens on the black market, particularly in the United States of America where the species is listed under the ESA and legal possession of one requires a permit. It was as a result of the black market that Bristol Zoo first acquired its four dragonfish, when they were found by HM Customs amongst a

fish smuggler's haul and given to the zoo.

The fish are notoriously difficult to breed in captivity. Previously the smallest body of water in which they had bredwas in a pond 5.5m by 5.5m and 1.1m deep. If the water quality and pH is not exactly right the fish, who are mouth-brooders, will usually swallow their offspring. The successful breeding in Bristol Zoo's aquarium tanks is thought to have been helped by raising the water temperature by several degrees and the use of reverse osmosis to ensure the water was clean as well as carefully monitoring the pH level.

Arowanas, also known as aruanas or arawanas are freshwater bony fish of the family Osteoglossidae, sometimes known as "bonytongues".

In this family of fishes, the head is bony and the elongate body is covered by large, heavy scales, with a mosaic pattern of canals. The dorsal and the anal fins have soft rays and are long based, while the pectoral and ventral fins are small. The name "bonytongues" is derived from a toothed bone on the floor of the mouth, the "tongue", equipped with teeth that bite against teeth on the roof of the mouth. The fish can obtain oxygen from air by sucking it into the swim bladder, which is lined with capillaries like lung tissue.

Chester Small Part Of Africa

Chester Zoo has unveiled an ambitious plan that, should it come to fruition will turn the zoo into the largest animal conservation, visitor attraction in Europe. The project that is estimated to cost a whopping £225 million in total will cover 80 hectares in addition to the 50 hectares currently occupied by the zoo.

The first phase of the planned development will be the creation of the UK's only complete ecosystem biodome, which will be called *'The Heart of Africa'*. The zoo is also seeking planning permission for a 90 bed hotel and a conservation collage, as well as revamping its main entrance and linking the site to a marina development on zoo owned land by the Shropshire Union Canal.

The *Heart Of Africa* will be based on an African rainforest ecosystem and house gorillas, chimpanzees, okapi, and a wide range of invertebrates, amphibians, reptiles and birds which will be able to roam freely amongst the vegetation. Visitors will explore the biodomes via canopy walkways and an interactive water 'ride'. Work is expected to be completed by 2018.

The Director General of Chester Zoo, Prof Gordon McGregor Reid, said:

"Chester Zoo already actively supports and runs conservation, science and veterinary projects in all corners of the globe but the increased visitor footfall from this new project will dramatically increase what we are able to do and fund out in the field. New and exciting animal species will be introduced.

This project is unlike anything that has been in the UK and will offer a spectacular visitor experience and the highest standards of animal welfare.

"Our intention now is to move forward to the planning stage and we will be working closely with Government agencies and our neighbouring communities during this process."

A lick of Paignton

Paignton Zoo and Living Coasts in Torquay in South Devon have also announced plans for improvement. The zoo's plans are less ambitious than Chester Zoo's but will cost considerably less money, £150,000, and most of the projects will be completed within the next year.

Paignton Zoo is transforming a former onsite educational centre into an 'amphibian ark' as part of an international campaign to help protect the many species of frogs, toads and salamanders facing extinction. A new zebra house will be built and a new enclosure added to the £1.1 million money heights to house a mixed collection of white-faced saki monkeys, goeldi monkeys and pygmy marmosets. New animals coming to the zoo include bearcats, African mongooses and porcupines.

Living Coasts will be unveiling the latest stage of its underground development and the zoo's new seahorse exhibit. In January this year the Marconi penguins were moved into a new purpose built enclosure at Living Coasts and their old enclosure will be used to house two South American fir seals from Hamburg Zoo.

Gaza Zoo

Whatever one may think about Israel's latest incursion into the Gaza Strip, there is one thing everyone should be able to agree on, the animals in Gaza Zoo should not have to suffer as a result of man's conflicts.In a good example to the governments of both Israel and Gaza an Israeli charity 'Let the animals live' has banded together with a Palestinian animal welfare organisation to get food and medical supplies to the animals of Gaza Zoo. Although they have managed to get permission to allow 30 trucks to go to the zoo the Israeli government wants to charge an additional fee of up to $350 per truck.

Eti Altman, the spokeswoman of Let the Animals Live said: *"The collaboration between us and the Palestinians is proof that the animals are not part of the political conflict, and anyone with a bit of humanity left in them, should volunteer and help out.*

"I thank all the organizations and the people involved who contributed to this important project of helping out the animals in Gaza who suffer hunger and diseases. In light of this humanitarian effort I have no doubt we can save many of the animals in the place. I am hoping that through the animals we will be able to draw the two sides closer together."

CRYPTO

Most people know that The Centre for Fortean Zoology, the parent organisation for CFZ Press, the publishing company responsible for this magazine, and the CFZ Alliance who organise Outreach Programmes and Direct Action Campaigns is the world's largest (and we like to think best) organisation dedicated to the search for new species of animal.

We have spent the last seventeen years working to bring cryptozoology into the mainstream, and for it to be seen as just another branch of the natural sciences, rather than something weird and wonderful allied to the study of UFOs or ghosts.

So, we make no apologies for including a section on cryptozoology in this magazine, because in many ways cryptozoology is the cutting edge of the study of the other animals who share the world with us.

Four Frogs Found On Flora Fauna Foray

In December 2008 four new species of frog were discovered in Cambodia by Flora & Fauna International.

Among the new species was Samkos bush frog (*Chiromantis samkosensis*) which, according to the team that found it has green blood and turquoise coloured bones. The unusual colouring of the frog comes from bilverdin, which is usually processed by the bodies liver, but passes back into the blood of this species. The bilverdin is actually advantageous to these frogs because they have thin, translucent skin so the greenish hue of their bones and blood helps to camouflage the frogs.

The frog appears to be very rare and it has been speculated that this is because the build up of bilverdin in the frogs bodies shortens the creature's potential lifespan.

Fauna & Flora International consultant naturalist and photographer Jeremy Holden, who discovered the Samkos bush frog, said:

"When I found the frog, I had a thrilling suspicion that we were looking at an entirely new species of amphibian. Photographing these frogs has been a challenge. They were extremely difficult to find, but thanks to their distinctive calls we managed to get some excellent shots and record them for posterity."

The Samkos bush frog was first spotted in 2000 but was only recognised as a new species as the result of this expedition.

The other new species of frog found on the expedition were the aural horned frog (*Megophrys auralensis*), the cardamom bush frog (*Philautus cardamonus*) and Smith's frog (*Rana faber*).

YEAR OF THE FROG

The charity *Amphibian Ark* declared 2008 to be `The Year of the Frog`, and mounted a twelve months campaign to raise awareness about the burgeoning crisis amongst the world's amphibia.

However, in a different way 2009 is proving itself more deserving of the title because as well as the four frogs described in the above story which were announced at the end of 2008, there have been ten new species from Columbia, and a whopping twelve from India.

We wonder how many more species will turn up this year.

CONTACT

Froglife
9 Swan Court
Cygnet Park
Hampton
Peterborough
PE7 8GX

Google It

In 2005 scientists from Kew Gardens in London were looking for an isolated area of African forest to which they could go, to find new species of plants from which they could collect seeds, to conserve species where rapid development is taking place.

The scientists used Google maps to search for an area of untouched mountain forest in Mozambique and found the uncharted forests of Mount Mabu.

Between October and November 2008 a team of 28 scientists and support staff hiked into the forest. The expedition leader, Kew botanist Jonathan Timberlake, said:

"The phenomenal diversity is just mind-boggling: seeing how things are adapted to little niches, to me this is the incredible thing. Even today we cannot say we know all of the world's key areas for biodiversity - there are still new ones to discover."

"This is potentially the biggest area of medium-altitude forest I'm aware of in southern Africa, yet it was not on the map, and most Mozambicans would not have even recog-

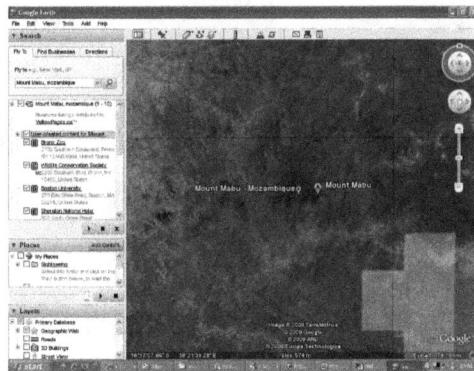

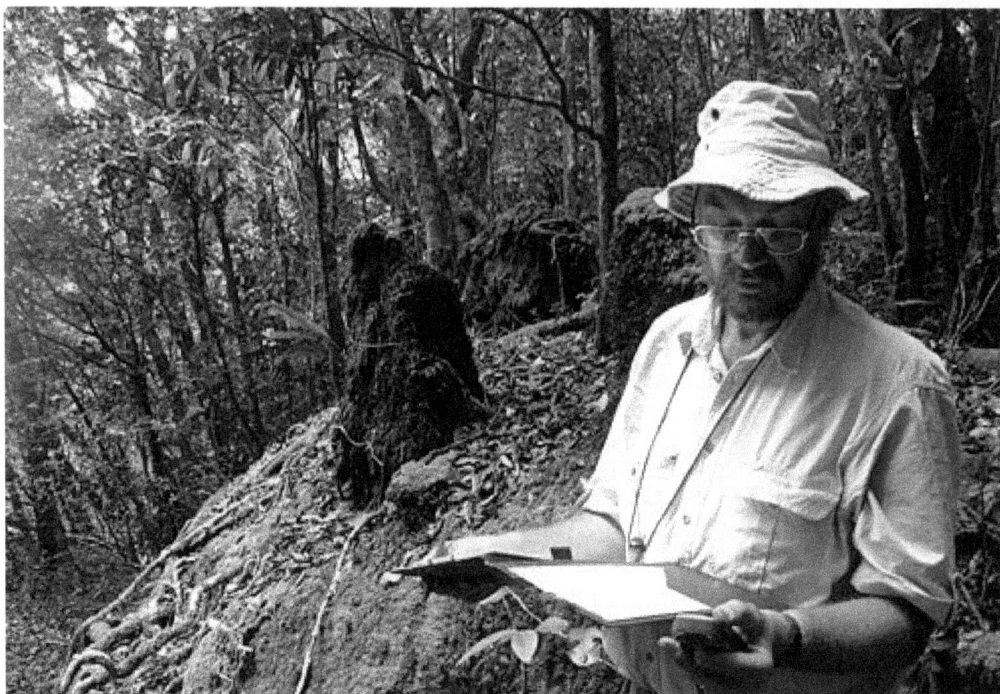

Kew botanist Jonathan Timberlake recording vegetation on the steep slopes of Mount Mabu. Photo: Tom Timberlake

nised the name Mount Mabu. Kew is working with the Mozambique government to protect areas like Mount Mabu and encourage local people to value the forest for its wildlife. By conserving the plant life we can help secure a future for all the other creatures we saw there".

Among the new animal species the team found were 3 new species of butterfly and a new species of adder.

The team also brought back 500 plant specimens that they will be studying and attempting to grow at Kew Gardens.

A new atheris viper

Carnivorous Sea Squirt

In January 2009 a research team investigating the impact of ocean acidification on deep sea corals off the coast of Tazmania discovered a new species of carnivorous sea squirt. The sea squirt is over half a meter tall and was found in water more than 2000 metres deep near to the Tazman fracture zone, a near vertical drop in the sea-bed between 2000-4000 meters deep.

Extinct pygmy tarsier rediscovered

In August 2008, a research team led by Sharon Gursky-Doyen of Texas A&M University had a welcome encounter with pygmy tarsiers in Indonesia. The species was thought to be extinct because they had last been observed by scientists in 1921, and the team is thought to be the first people to see the animals alive since then. The expedition was launched after a pygmy tarsier was accidentally caught and killed in a trap two Indone-

sian scientists had set to catch rats in the Sulswesi highlands in 2000.

The A&M team used 276 mist nets in their two month long attempt to capture live pygmy tasiers that they could fit with radio tracking equipment and managed to catch two males and one female.

The creatures were observed before being released and Gursky-Doyen observed that the tarsiers all had claws instead of fingernails, making the pygmy tarsier unique among primates.

Gursky-Doyen believes that the claws are an adaption to their environment as these help the tarsier to hang onto the forest's moss covered trees and branches.

Gusky-Doyen believes that studies like this prove that there are still many species of primate waiting to be discovered in the world.

Eggs!

One year ago, Ivan Ineich, a herpetologist with the Natural History Museum of Paris, collected a single gecko egg while on island of Espiritu Santo in Vanuatu. Ineich took the gecko egg back with him to Paris where it hatched and he waited until it was fully grown before he declared it to be a previously undocumented species.

The species is not thought to be rare in Vanuatu where the geckos live in, or near, plants that are infested with ants as high as 20 metres high on trees.

The Emirates Natural History Group.

Founded in 1977 by JNB "Bish" Brown this active group, which has chapters in Abu Dhabi, Dubai and Al Ain, is a vital force, speaking up for the environment in one of the most un-ecological countries in the world.

As well as the advisory roles that some members take on, the group has a lot to offer the amateur naturalist with regular field trips investigating the various ecosystems of the Emirates and Northern Oman. Opportunities to do such things as live trapping of small mammals and light trapping insects are part of the regular activities arranged by the group.

The group actively encourages the enthusiasms of the members whether they are professional ecologists or just nature lovers. There are regular lectures on various aspects of Natural History, as an example, there have been lectures on long horn beetles, Socatra Island, and the importance of Taxonomy. This last subject is indicative of the vital work that the group undertakes in surveying and categorising the wildlife of the region.

A special interest is the insect and arthropod life of the UAE and a collection of specimens is being built up and curated. This is the life project of one of the members, who also had the good luck to come across a very unusual species of Picturewing fly. Findings such as this are reported in the group's Journal 'Tribulus' which is published twice a year in both English and Arabic as well as in the more informal monthly newsletter and on the website.

Heather Mikhail

www.enhg.org
http://groups.yahoo.com/group/ENHG/

The British Killifish Association

The British Killifish Association is dedicated to the study of, propagation of and publication of knowledge pertaining to Killifish.

What are the benefits of membership?

If you join the British Killifish Association you will receive our all colour monthly journal 'Killi-News'. This journal contains articles on all aspects of killi-keeping together with descriptions of new species, collecting trip reports, a Fish and Egg List which provides a forum for members to trade livestock, and information on local area meetings (some of which are members only), auctions and our convention.

If you would like a free copy of Killi-News please contact the registraar: Brian Wilson. You will also receive:

- BKA starter booklet 'A Guide to Killie Keeping' by Fred Wright; covering killie maintenance, feeding, breeding, diseases and posting fish.
- Nominal role; a list of names and addresses of all BKA members, so that you can contact members nearby. You should also be supplied with names of members willing to answer your questions.
- Rules and Services guide; details of association rules, and the BKA merchandise listings.
- Access to the BKA website members area which contains a whole lot more pictures and articles.

Brian Wilson
Gordleton Farm
Silver Street
LYMINGTON
Hant's
SO41 6DJ

CLUB NEWS

Ryedale Aquarist Society

**10th Open Show & Auction
Sunday 26th April, 2009**

at

**Kirkbymoorside Memorial Hall,
Market Place, Kirkbymoorside,
nr Pickering, North Yorkshire
Y062 6AD**

Incorporating a round of The British Livebearer Association Fancy Guppy League

**Open Show exhibits - 20p
Guppy League exhibits – 20p**

Schedule of Classes

1. Danio and Minnow
2. Rasbora
3. Guppy
4. AV Female Livebearer
5. Molly
6. Platy
7. Swordtail
8. A.O.V. Livebearer
9. Breeders - Livebearer
10. Matched Pair - Livebearer
11. Polypterus (Birchirs)
12. Betta splendens (Siamese Fighters)
13. Ctenopoma and Microctenopoma (Bushfish)
14. A.O.V. Small Anabantid (up to 10cm)
15. A.O.V. Large Anabantid (over 10cm)
16. Small Barb (up to 10cm)
17. Large Barb (over 10cm)
18. Angel and Discus
19. Endemic Rift Lake Cichlids
20. A.O.V. Small Cichlids (up to 10cm)
21. A.O.V. Large Cichlids (over 10cm)
22. Small Characin (up to 7cm)
23. Large Characin (over 7cm)
24. Junior
25. Sharks and Foxes
26. Fundulopanchax, Scriptaphyosemion, Archiaphyosemion, Aphyosemion and Callopanchax
27. A.O.V. Killifish
28. Botia (Botia, Leptobotia, Sinibotia, Parabotia, Syncrossus, Yasuhikotakia, and Chromobotia)
29. Loach
30. Single Tailed Goldfish
31. Double Tailed Goldfish
32. A.O.V. Coldwater
33. Rainbowfish
34. True Albino
35. Aspidoras and Small Corydoras (up to 5.5cm)
36. Scleromystax, Brochis and Large Corydoras (over 5.5cm)
37. Synodontis
38. Loricariidae Catfish (Armoured Suckermouths)
39. A.O.V. Small Catfish (up to 15cm)
40. A.O.V. Large Catfish (over 15cm)
41. A.O.V. Tropical (up to 15cm)
42. A.O.V. Tropical (over 15cm)
43. Matched Pair - Egglayer
44. Breeders - Egglayers
45. A.V. Fish from the African Continent.

**OPEN SHOW AWARDS;-
Y.A.A.S. AWARDS
BEST IN SHOW
BEST EXHIBIT**

BEST GOLDFISH EXHIBIT

PRIZES FOR: 1st 2nd AND 3rd PLACE IN ALL CLASSES

CLUB NEWS

JUDGES
Mrs. P. Jones
Mr. T. Douglas
Mr. S. Grant
Mr. K. Webb

THE B.L.A. FANCY GUPPY LEAGUE WILL
BE OPEN TO B.L.A. MEMBERS ONLY.
JUDGES, CARDS AND PRIZES ETC. WILL BE
PROVIDED BY THE B.L.A. JUDGING TO
IKGH RULES AND STANDARDS. SHOW
TANKS TO BE 7"WIDE, 7" HIGH AND 3"
DEEP.

The 3 B.L.A. classes for the day are as follows: -
1. Male Short Tail Guppy.
2. Male Broad Tail Guppy.
3. Male Sword Tail Guppy

BENCHING OF OPEN SHOW AND B.L.A. EX-
HIBITS

10.00a.m. to 1.00p.m.

AUCTION
LOTS TO BE TABLED BY 12.15p.m.
TO PRE-BOOK AUCTION LOTS PHONE
01751 472715

AUCTION BEGINS AT 12.30p.m.

AUCTIONEER: Mr. S. Jones

Please note that Ryedale A.S. will take a 15%
commission on all auction lots sold.
Please label lots to assist the auctioneer.

Please make sure that all electrical goods are
clearly marked with the name and address of
the seller.

Exchange Publications

We welcome an exchange of publications
with other like minded periodicals of mu-
tual interest.

We currently exchange with:

Killi-News
Contact details see page 24

The journal of the British Killifish Associ-
iation, is - as the name implies - concerned
with killifish. Issue 519 (December 2008)
features a fascinating description of the
discovery of four new West African species, together with an account of the me-
chanics of killifish speciation, and a blow-
by-blow account of building a fish room.

The Avicultural Magazine
c/o Paul Boulden
Arcadia
The Mount
East Allington
Devon TQ9 7QJ
Tel/Fax: 01548 521203
http://www.avisoc.co.uk

The magazine of the Avicultural Society,
which has been published since 1894. It
covers all aspects of bird keeping concerning non-domesticated species that is likely
to be of interest to aviculturists.

CLUB NEWS

Simon Wolstencroft
1958-2009

Simon Wolstencroft one-time editor *Tropical Fish* magazine and founding editor of *Tropical World* magazine has died of liver cancer. I am particularly sad because for many years we were very close friends.

I am certain that everyone reading this will know that there was bad blood between us at the end, and that there had been for several years, and I feel that it would be completely hypocritical of me to pretend that this never happened.

Iggy Tavares, bless him, had tried to mend the rift on several occasions, but it was too deep and too fundamental, and - given the circumstances - was never going to be fixed.

But I wanted to mark his passing, and remember the kind, funny, gentle, irreverent, and sweet man who was my friend for five or six years, before the illness took control..

When he was at his peak, he was the best editor that I ever had, and I learned a lot about the editor's craft from him.

After he and I parted company I missed him a lot, and always hoped that somehow we would become friends again, although I knew that this was never going to happen. I still miss him, and my heart goes out to Debbie at this sad time.

Despite everything, I learned a lot from him about how to, and how not to, run a magazine, and so it is with sadness that I dedicate this debut issue of *The Ameteur Naturalist* to him. Rest in peace you old bugger!

OBITUARY

THE CENTRE FOR FORTEAN ZOOLOGY
www.cfz.org.uk

WHICH SIDE ARE YOU ON?

OUR HERO GERALD DURRELL (1925-1995) WAS THE FATHER OF MODERN CONSERVATION. HE BELIEVED THAT CHILDREN SHOULD GROW UP SURROUNDED BY ANIMALS AND BOOKS.

WE AGREE WITH HIM.

BUT IT SEEMS THAT OTHERS DON'T!

CFZ OUTREACH: Educators with Attitude

COSTA RICAN ADVENTURE

by David Loft

After thirty hours of travelling and sitting in airports Alison and I finally arrived in Costa Rica, but first more travelling would be needed, as our final destination would be far from any towns.

We spent a further two hours travelling via taxi to the village Las Horquetas. This was where we spent our first day in Costa Rica, at the Sun Sun lodge. Here we spent our time looking around their stunning gardens, where we first saw what we were here for. Sitting on a wall bold as day was a mature male brown basilisk(*Basiliscus vittatus*) that looked as if he'd had his fair share of close calls or territorial battles, the latter being the most likely as we later discovered there was a pair in nearly every wild plantain we looked in. The garden was also in abundance of anoles and geckos in the trees, ameiva's running through the leaf litter and cane toads rustling in the bushes. Also, this was where I saw my first wild leaf cutting ant colony and was really impressed with the distances they travelled for food. So much in such a small area made us even more excited about getting to the premontane rainforest, 2,000 feet above sea level.

That night Alison and I had our dinner and headed straight for bed as the next morning we would have more travelling to do. This would be the final stretch and would be

Costa Rica, which translates literally as 'Rich Coast', is a country in Central America, bordered by Nicaragua to the north, Panama to the east and south, the Pacific Ocean to the west and south and the Caribbean Sea to the east. Costa Rica was the first country in the world to constitutionally abolish its army. Among Latin American countries, Costa Rica ranks 4th in terms of the 2007 Human Development Index. The country is ranked 5th in the world, and 1st among the Americas, in terms of the 2008 Environmental Performance Index. In 2007 the government of Costa Rica stated that they want Costa Rica to be the first country to become carbon neutral by 2021.

A male basilisk basking

Anolis species abound in this region

the toughest part of all. We awoke fairly early and our taxi arrived to take us to the next mode of transport which would be a tractor. Two hours we spent in this tractor trailer and it was not a pleasant experience, it was extremely bumpy, mud flying all over us, being thrown from left to right. One of the four tyres on the trailer punctured and came off completely, and it took five minutes for the driver to even notice. If this wasn't bad enough it was followed by a brisk hour trek up hill. This gave us our first taste of how tough the terrain is. Everything is wet with moss and leaf litter covering the floor. Slipping and falling is a regular occurrence. I was certainly thinking at this point that this best be worth it! The terrain is tough and unforgiving and so dense, but so beautiful and green. Moss and bromeliads smother the trees with the sounds of insects all around you, with the flash of a lizard or frog diving into the dense foliage - it was truly beautiful. A feeling came over me and I thought to myself "*How in hell will we find anything, let alone get near enough to handle or photograph it, this is going to be tough*". After being shown to the riverside cabin, we had our dinner and decided to leave searching for animals to the next day as we were exhausted from our two days of travelling. So, we headed to our cabin, but on the way couldn't help but scout for things.

When we arrived at the cabin we searched outside and underneath it to see what we would be sharing our accommodation with, this was a little unnerving as there was a two-inch scorpion, tailless whip scorpion and lots of large wolf spiders all over it which took advantage of the moths attracted to our torch lights. If this wasn't unnerving enough, we discovered a Wilson's

Red eyed treefrog

One of the giant mantids, probably a Stagmoptera species

Common boa up a tree

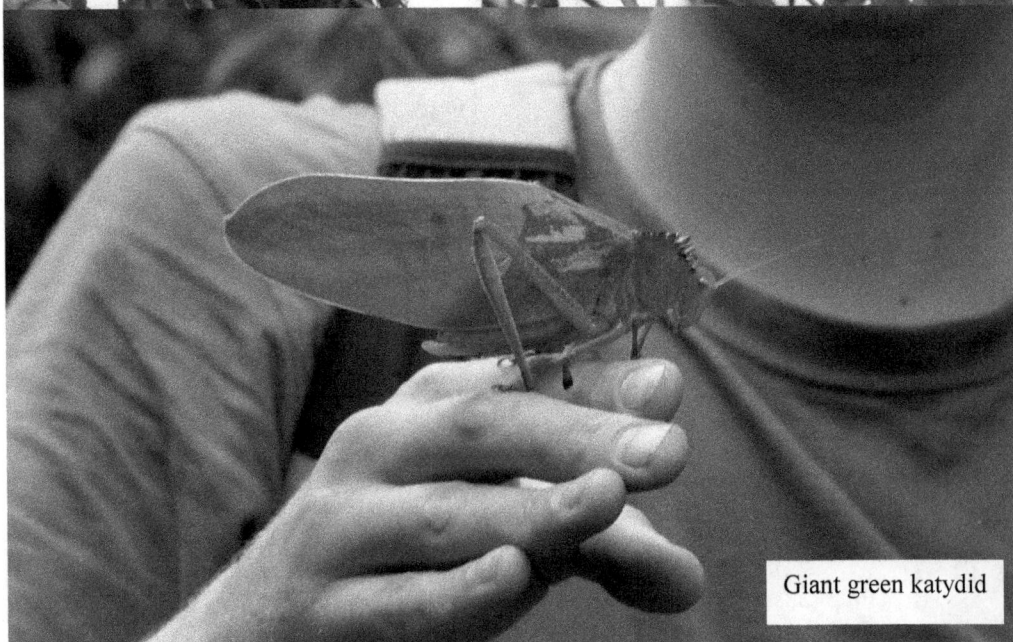
Giant green katydid

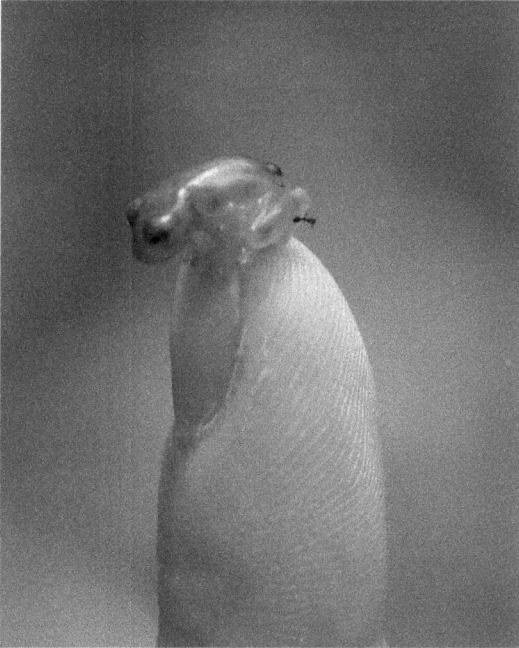

climbing rat had taken up residence in the wardrobe and believe me they can climb. He was also determined to get at our sweets and hot chocolate. Considering all this, we slept really well. Next morning, after enjoying a cold shower, we decided where to explore first over breakfast and agreed to head to the two thirty foot waterfalls only a ten minute trek away. On the way down the edge of the river, we searched for whatever we could find. We found a huge - what we believed to be a very large - katydid and there were plenty of stream anoles and the quick glimpse of a strawberry poison arrow frog. We were really hoping to find ourselves a snake, but we were out of luck. Not all was bad; the waterfalls were a sight not to miss; a real piece of paradise.

We headed back to our cabin a little disappointed that we hadn't discovered more, but with high hopes for the night walk we were going to take. You felt most vulnerable during the night walks as you got a real sense of being watched. A lot of the night walks would get cut short because this is when it would rain most heavily and make the already difficult trails almost impossible to navigate. We never gave in though and went out with head torches looking as soon as it would become calm. We found some stick insects, mantis, masked tree frogs, and red eyed tree frogs among other tree frogs. We also had a few regulars outside our cabin. There was the pink toe tarantula with her newly laid egg case under the cabin, also a slim fingered rain frog and a little glass frog.

Four days of constant searching for animals and still no snake. I must say I was starting to get frustrated, but our luck was about to change on a trail we hadn't tried. I was checking out a hole in the ground with a torch when Alison suddenly tells me to step back in a concerning tone. As I do, I ask what's up? To which she replies "eyelash" and as my eyes become focused, I realised

Two of the arachnids with whom we shared a hut.
ABOVE: Wolf spider BELOW: True bird eating spider

The amphibian diversity in Costa Rica is far greater than that of Europe and was almost bewildering for us British travellers to behold.

that I just had my face one foot away from an eyelash viper (*Bothriechis sclegelii*). This was an amazing sight and was at the top of our list of animals we wanted to find. That evening I read that one of the most common places to get bitten by eyelash vipers is in the face. I must say I was fairly happy I didn't add to the ninety to a hundred reports of bites recorded by this snake each year in Costa Rica. All frustration had disappeared at that point until the next day looking for snakes all morning, without any luck I spotted a bird of prey swoop down in a clearing. Alison and I both stared as the bird hit the ground, I shouted "it's got something" and we proceeded to get closer, when we realised it had the very thing we had been searching for - it had a snake! We both gasped. We later read that the bird was a laughing falcon, a snake-eating specialist that would wait in a tree scanning the surrounding area for a snake to come into sight. It then swoops down and grasps the snake in its talons, and then flies back to the safety of the trees and decapitates and eats it.

After a bird had just eaten the very thing I was looking for I got a sudden determination and headed straight back out searching and I didn't fail. Out of pure luck, and my newfound determination, I spotted a five foot common boa in a tree. There was only one problem. This tree overhung the rocks in the river, covered in moss and, even worse, extremely fierce and easily irritated bullet ants. After a quick risk assessment we decided to leave it be. But I did find a snake!

Our time in the rainforest had come to an end and it was our last night. This went out with a real bang. No we didn't have a party. Thunder and lightning!. The thunder was so loud and powerful our whole cabin shook and felt as if it would collapse any minute. The rain was heavier than anything I have ever experienced, so much that you would be soaked through within seconds. As quickly as it came it was gone.

This trip for us was a real experience of a lifetime. We arrived naïve and were expecting animals to be leaping out from all over, but soon learned otherwise. It was tiring and hard work but we wouldn't change a bit, we only wish we could have stayed longer.

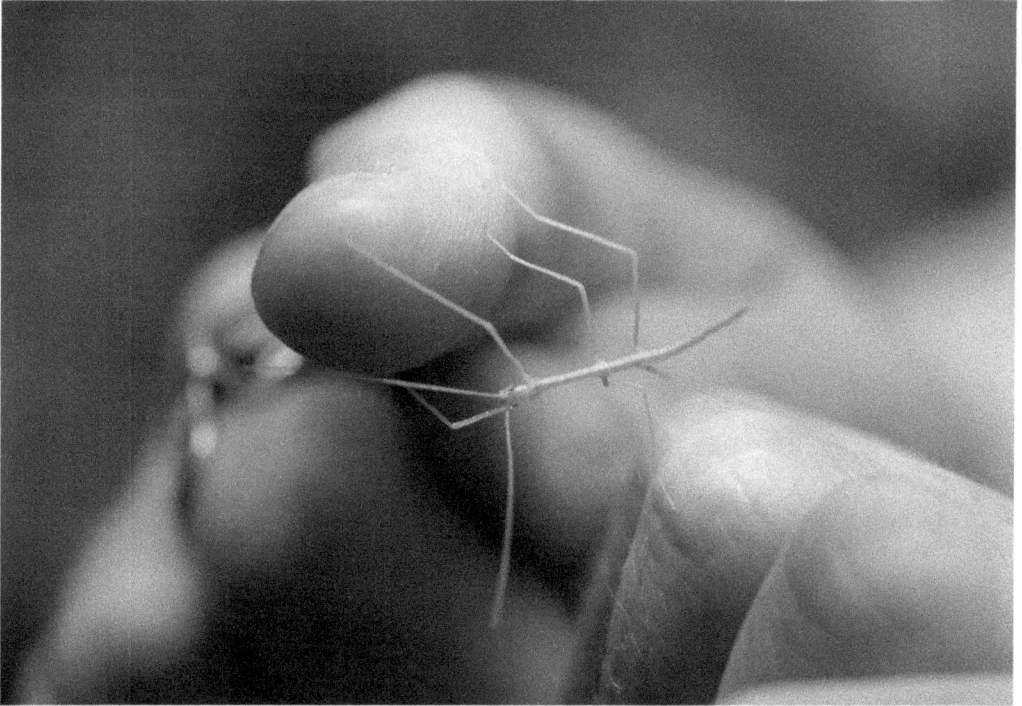

Fruit-Beetles with Attitude

by Max Blake

Beetles from the sub-family Cetoniinae have been popular subjects for the Exotic Pet trade for years thanks to their easy and cheap maintenance, rapid breeding cycles, activity (well, as adults anyway!) and stunning colouration. *Eudicella* species are amongst the most popular. They come from the Ivory Coast, round to Ethiopia and Namibia. They belong to the Tribe Goliathini, which puts them close to *Chelorrhina*, *Dicronorhina* and the giants among giants, *Goliathus*. *Eudicella* males are larger than females, with a large bi-furcating cephalic horn. Their first pair of legs are also longer which help them to grip onto females when mating, plus they have a depression on their underside which slots onto the female's wing case to help the male mate.

There are a number of species of *Eudicella*, but the actual amount varies depending on who you talk to as a number of species have been called sub-species, some are colour variations and others which are species are called colour variations, making the whole thing very complicated indeed! There are roughly between

20-30. In general, this guide to keeping them will be fine for all the species commonly encountered in the hobby, including *E. gralli*, *E. smithi* and *E. trilineata*, amongst others. Most species grow to between 30-50mm.

They do best with substrate made up either by yourself or by a breeder, with about 10% white or brown well rotten wood, broken up into very very small particles (remember, the smaller the particles, the easier it is for the larvae to eat and so the easier it is for it to get on and eat more, helping it grow both larger and faster), and the rest made up of a 50/50 mix of well broken up beech or oak leaves, at least 3 years old, organic compost or peat (this must have been left outside for a year or so before hand and open to the elements to allow the chemicals in it to break down as so they don't harm the beetle larvae). Failing that, using old substrate made up mainly of frass (the larvae's excretory products) will do as an alternative. Indeed, it is usually good practice to add a handful of old frass from the same or a related species as this contains some bacteria which help to continue to break down the food in the beetle larvae if it eats the frass. The symbiotic bacteria living within the larvae's gut help to digest the tough cell walls which plants have, allowing the larvae to extract more nutrients. The substrate must be kept damp at all times, just damp enough so that it holds together when you hold some in your hand to compress it.

You can add a supplement food if you want too. This speeds up growth and makes for larger adults. Eudicella accept apple chunks, dog kibbles and cow dung as extra food. The apple chunks should be placed into the substrate and removed after a day to prevent rot; the dog kibbles should be of the soft kind (a well known brand produces theirs with gravy which are ideal), broken up into eighths and removed after two days, and cow dung needs to be fully dried out and crumbled into the substrate. This is a good addition, but after the week or two it takes to dry it is usually smell free!

Eudicella can and do cannibalise, but not as badly as some of their larger relatives. A rough guide if you want to keep them together is to allow about 1ltr of substrate per larvae (*E. gralli* is a notorious cannibalizer, so it is best to allow 1.5-2ltr for them). It has been suggested that adult females will eat young larvae, but in my experience as well as others who keep these beetles, there have been no problems. They generally take 4-8 months to grow from egg to adulthood, with the adult living 2-4 months. The pupal cells are formed from the substrate, and are usually fairly tough. They are best ignored, but placed on top of the substrate so you can see when the adult beetle hatches. One habit *Eudicella* have is that they often build the cell cemented to the tub they are in. If this happens, do not disturb it! If you do it makes it very easy for other larvae to enter and eat the pupae. If you do accidentally break it off, take it out and place it on top of the surface with the hole facing upwards. It is quite nice if they do build it on the side as you can see the transformation from larvae, to pre-pupa, pupa and then adult. You can usually see the adult hardening up in the cell; it hardens and darkens in colour until it emerges.

To breed the adults, you need a ratio or about 1 male to 2-3 females for the best results. This makes sure that all the females are fertilised, and you still get a good batch of eggs. A tank of about 10 gallons is fine to breed 3 males with 6 females. This means that (as long as 2/3 full with substrate) that the beetles have a lot of space to move, feed and fly in, plus lots of substrate to lay in. The mix described above for raising the larvae will be fine for larvae. The adults are best fed on beetle jelly or soft fruits like mango or banana which should be replaced before it rots. The females will spend a lot of time underground laying their eggs, but the males are usually out and about; their horn makes digging very hard! The eggs are usually fairly obvious as little pale balls in the substrate. They can be left in the substrate to develop. Young larvae (L1) can be removed and reared separately. They are best kept at between 20-28 degrees Celsius, so room temperature usually poses no problems.

So there we have it, a simple guide to keeping some of the most energetic beetles around. If you are undecided about keeping them, just look at the male's horns. The sight of one of them zooming around the room bumping into every object in sight is something not to be missed!

TALES FROM THE BUSH: In the footsteps of Wallace

by Steve Backshall

Drifting along in the pitch black a million miles from the modern world, our eyes suddenly wake to something totally unexpected. There is not supposed to be any human habitation for several days, yet the rainforest river banks have inexplicably been lined with Christmas trees, draped with dense curtains of blinking white fairy lights. The tiny bulbs cast dramatic flickering shadows over the tangles of the forest canopy, and shimmer off the water - unnatural light in the heart of darkness. Even from across the mighty Siretsj River, the glow is so powerful that I could read a map by it.

If only there was a map to this forgotten place. It's impossible to describe the eeriness of finding a lit town there in the midst of the wildest forests of the world; you'd expect us to be desperate to throw ourselves into the arms of human company, but instead it's actually deeply threatening. My first instinct is to huddle down into the water dregs slopping around in the bottom of the dugout, and avoid the prying eyes from this bizarre settlement at the end of the world. Who knows what our reception would be - spears and arrows, perhaps worse. The hairs erect on my forearms, I hold my breath so my exhalations don't give us away. However I should have known from that silence - the absence of generators, radios and televisions and any noise other than the burping of frogs and cheeping of bush crickets - that this was not the ghost town it appeared to be. As

Steve Backshall is a remarkable bloke. Whilst in a body cast, recovering from a broken back he looks back at his adventures on the island of New Guinea… Under the old magazine this was originally going to be a two part article, and indeed we printed part one in *Exotic Pets #6*.

However, at the time I felt that it was a crime to edit it, and now that space constraints are no longer an issue, I make no apologies whatsoever for having repeated what we printed last time, and presenting the entire article for you uncut… JD

Alfred Russel Wallace, OM, FRS (8 January 1823 – 7 November 1913) was a British naturalist, explorer, geographer, anthropologist and biologist. He is best known for independently proposing a theory of natural selection which prompted Charles Darwin to publish his own theory.

the boat drifts up to the other bank and the blinking lights come up on our bow, the billions of fireflies all become clearly visible, making a Mardi Gras of the hostile forest. There are no people here, just an infinity of tiny insect miracles all searching for their perfect partner. They cover the bushes no more than inches apart, and for hundreds of metres along the banks, glowing and flickering in unison with those around them. The greenish white bioluminescence is created from a mixture of chemicals inside the abdomen, and is about the purest form of light known to science - formed with no ultraviolet or infrared rays and almost totally without heat. The main enzyme at work is luciferase - a reference to the fires of hell and the demon that unleashes them. Though I've seen fireflies and glow-worms all round the world, nothing has come even close to this synchronised display illuminating the dead dark of the forests, it is quite simply the most awe-inspiring invertebrate spectacle I have ever seen. Rapt fascination is abruptly and sickeningly broken, by a SWOOSH in the water and a violent splash that soaks our already shivering forms, a huge beast wrenches its prehistoric bulk from the water just metres off the side of the boat.

There is a flash of eyes, lit up like hot coals in the glancing light of my torchlight, and a mighty tail disappears leaving a churned maelstrom behind. The crocodile must have been bigger than our boat, but he vanishes instantly into the ink black of the river deeps. The jokes about our tiny dugout's flimsiness and instability are suddenly not funny anymore. I have never felt so alone, so vulnerable and so very far from home.

It is the mid 1990s, and the vast jungles of Irian Jaya were only just beginning to be exploited, and held biological riches beyond my wildest dreams - this was not just one of the world's great hotspots of biodiversity with millions of species of exotic invertebrates and bizarre mammals, but was one of the most important places in the world for the history of the biological sciences. It was the great zoologist Alfred Russell Wallace who was the first to notice that as you travel east through Indonesia, the flora and fauna changes dramatically, with a big change around about the middle of Indonesia. The region where the crossover occurs has been dubbed Wallacea in his honour, with the very basic premise being that New Guinea once shared a land bridge with Oz, (maybe as close as the last ice age) whilst Java, Sumatra, Borneo and Bali were once connected to Asia. Even today the gulf between the two small and comparatively close islands of Bali and Lombok is extraordinarily deep, and travelling between them you feel as if you are leaping from one continent to another. The vestigial wildlife is a result of this ancient connection; from distinctly Asian in the west with large mammals such as tigers, rhinos and elephants, to distinctly Australasian in the east with marsupials and monotremes, that is mammals that lay eggs!

New Guinea was, to my hero Wallace, what the Galapagos Islands were to Charles Darwin. A place where the greater picture of how the world's species had developed became increasingly evident, and the grand idea of evolution became an uncomfortable, but unavoidable truth. Wallace spent most of the 1850s and 1860s travelling through eastern Indonesia, and lived for several years with the natives in Papua

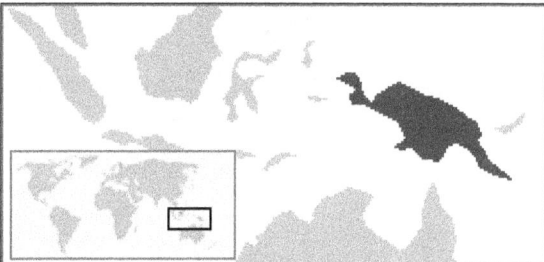

New Guinea. After all those years studying the wildlife of the region, Wallace was finally struck with the delirious epiphany of the "survival of the fittest" whilst in a malarial fever in the Spice Islands. These ideas predated those of Darwin. In fact, Darwin was probably several years from publishing his theory when Wallace sent him his own thesis, hoping for advice from the older man. The fear of being gazumped was apparently what spurred Darwin into publishing his first papers on the origin of the species, and bringing perhaps the greatest idea of all time into the public domain. This aspect alone of his life, his astonishing but overshadowed affect on science is enough to make Wallace my fondest hero. However, it was a fascination with his travel writings that first drew me to New Guinea. Though you couldn't dispute the genius of Darwin, his journals are all rather dry, with little flavour of the places he visited, and little sense of the adventures he must have endured.

Wallace's books on the other hand had more swashbuckling and Derring Do than Ryder Haggard and Wilbur Smith's back catalogues put together. This was a man who in the late nineteenth century put his life in the hands of headhunters and cannibals, discovered new wondrous species of animal every day of his travels, and hacked his way through terrain that, even today, is some of the most thankless on the planet. I would give anything to have been with him on those journeys, but as this was unlikely to happen outside of my dreams, my only option was to follow in his footsteps.

An airborne hour was spent looking down on nothing but lightly undulating oceans of jungle, cut through with vast, meandering brown rivers and not so much as a glimmer of visible life. After landing, however, it was a completely different story. Senggo is the site of the largest landing field in the whole of the vast Asmat area. The village consisted of several rickety death-trap plank ways teetering over dark green ooze. There were a few wooden shacks, which could well have been knocked up in a bored afternoon, some mangy dogs scratching their ticks, and a few muscular locals moping around in the remaining threads of T-shirts and shorts that had been imposed on them by missionaries years before, and had obviously never been taken off.

Any wildlife of even vaguely edible size around Senggo inevitably found its way into the pot, but the ironwood struts of the boardwalks provided a great substrate for giant golden orb weaving Nephila spiders to make their fabulous webs. These spiders are an arachnophobe's worst nightmare, being the size of your hand with long scuttly legs. The females look like giant monsters, sat in the middle of webs that can be the size of a table tennis table. The threads are so strong that I've had to rescue bats and even small birds from them before, and have managed to dangle a video camera (very carefully!) from a collection of silken

strands. It's one of my favourite tricks to take one of these evil looking beasts from a web and let it wander over my hands - none has ever bitten me in years of trying, though they look as if they could kill you just by looking at them. In many parts of the archipelago, kids will collect Nephila webs round sticks, and use them for fishing - tiny fish going for the insects caught in the threads, and getting stuck to the gooey strands.

The squalor and ugliness of Senggo village was soon forgotten a few canoe hours from the grass airstrip. Nervous crocodiles were ducking beneath the surface at our approach, there were streaks of colour in the forest telling of fabulous birds, and reptiles dripped from mossy boughs. Before us in the river, a skinny periscope with a beak, popped out of the water to survey us; a snake!

Well, no, a body and then wings emerged from beneath the water's surface and the 'snake-necked bird' or darter climbed up on to a log. Somewhat cormorant like, these incredible birds allow their feathers to become waterlogged so they can swim around submerged with just their neck and head poking up from beneath the water. This allows them to hunt underwater without all the air bubbles in the feathers encumbering them, spearing fish with their harpoon beak. Of course, sodden feathers are no use for flying, so you're most likely to see these birds stood on riverside branches with their wings extended, drying out in the sun.

Back paddling down the muddy river, and a flash of green iridescent feathers passed in a bouncing flight mere metres from our canoe. My guide became wildly excited and desperately ditched his paddle to scrabble around for his spear, rocking the flimsy dugout and threatening to plunge us all into the mud. "Cenderawasih" he shouted - 'Bird of paradise'. But it was already gone. This was to be my only view in the wild of these mythical birds on this trip. As a child, stories of the Birds of Paradise were what drew me to Papua New Guinea above any other destination on earth. The allure of vast, dark rainforests inhabited by jewelled birds like flying angels my most romantic vision. Unfortunately, they are a sight that my children will probably never have a chance to see. There are 42 species, 36 of which live only here. The birds themselves have been hunted constantly over the last 9000 odd years since people first made their way to New Guinea, but low human populations and the immense areas with no people at all, meant the birds could always find pockets of forest to live in peace. However, in the late 1800s the couturiers of Europe discovered the exquisite feathers, and began a trade in them for hats and other adornments, that was to see hundreds of thousands of pelts transported across the world. The trade in the birds has continued in various guises right up until the present day with the birds obviously being the number one prize for any bird collector. All these trades may be heavily regulated now, but for native New Guinean peoples the birds have always been used to adorn their headdresses and supplement traditional bride prices; how do we tell them, that because of our actions they can't take them anymore? The birds themselves are a victim of evolution; handicapped by their lush feathers to be slow and blatantly obvious in the forest dark. They also display for their females in the same places annually - often clearing patches in the forest canopy so that sunbeams will spotlight them on their dancing stages. They couldn't be easier to catch if they fell out of the sky on to village dinner tables ready-stuffed. Add to that the Indonesian government's shameful strip mining of the Irianese forests, and their conservation will soon be out of our hands. The birds of paradise that remain are the last of what may already be a relic population and cannot possibly survive.

Whilst my sighting of the famous Cenderawasih was very limited, a little later, we got a very good look at another of Irian's airborne wonders, the mambruk or Victoria crowned pigeon, the largest, and by far the most beautiful pigeon in the world. The first one we saw seemed to be showing off for us, flying ahead of us down the river to perch in a tree until we arrived, before flying on ahead again as we drew near. We saw the bird in two different colour morphs; jet black with blood red trim, and vibrant indigo with dark blue and white. Both had bright red eyeballs. The mambruk also has on top of its head a spray of feathers, like the seedlings of an autumn dandelion.

There were kingfishers in the mangroves, and a variety of waders in the river mud, which also sparkled with colourful fiddler crabs and spluttering guppies. The most conspicuous bird was the hornbill, with the gaudy colours of its face and a hollow horn topping its toucan-esque beak. The arrival of this mammoth beauty was always announced by the majestic sweeping sound of its wings, like the swoop of slow-motion helicopter blades, audible when the bird was still well out of sight.

Some of the most dramatic creatures however of Papua are marsupials. Tree kangaroos and cuscus, feather tail possums and bandicoots. Many marsupials live on plant sap, nectar or insects but the quoll is a particularly vicious little predator that feeds on birds, reptiles and small mammals. About the size of a domestic cat and with big dark eyes, its exceedingly cute looks hide a ferocious demeanour and savage predatory instinct.

Irianese fauna find their place in the Guinness book of Records in the shape of the Bandegerou or Mbaiso whistling tree kangaroo. This black and white species of tree living kangaroo emits a characteristic whistle, measures about four foot long and weighs fifteen kilogrammes. It was discovered in 1994 by one of my great heroes Dr Tim Flannery, making it one of the largest new mammals to be found in recent times. The world's only known poisonous bird is also found here; the New Guinea pitohi, whose feathers and flesh contain one of the most powerful toxins around, poached from berries that are its main food source and used to deter predators. Irian is also the possible home of the "most frequently seen extinct species" in the world, which if it does still exist, is the world's largest carnivorous marsupial. The thyalcine or Tasmanian tiger, is an incredible creature, whose most novel feature is the capacity to open its maw so wide that the upper and lower jaws form a single parallel trap of fierce teeth. During my visit in 1997, the newspapers were abounding with tales of a lone male having been spotted up near Jayapura on the boundary with New Guinea. Quite something, as the last recorded specimen died in Hobart Zoo in Tasmania in 1936. Tazzie remains totally obsessed with the creature, and people still wander its wilderness hoping for the unimaginable scientific kudos of finding living examples of the beast.

At late dusk, as I paddled upstream with my muscular Irianese oarfellows, one of the most incredible noises in nature began to soar around the arena created by the wall of trees that lined the river. An unsettling, high-pitched scream that wailed past me along the riverbank in a Mexican wave, before haring back past again like some ethereal police siren, racing into the distance. The volume was deafening; if we'd wanted to have aconversation we'd have had to scream our lungs out. This indescribable miracle of noise is made by millions of male cicadas in the trees, set off by dusk's fall. It is the loudest noise made by any invertebrate, and can exceed 110 decibels - about the same as an industrial band saw. Incredibly if you were working with this kind of noise, you would by law have to wear ear-protectors or face irreversible damage to your ears! On my return to London, I contacted an entomologist at the Natural History museum, and he told me that this noise is so precisely timed, that scientists recording the phenomena can set up their equipment mere minutes before they know it will begin.

This incredible noise is most remarkable when you actually get one of these cicadas in your hand. The abdomen is pretty much hollow, and is just used to amplify the sound created by the tympanums, two cymbal shaped organs where the armpits would be (if an insect had armpits!) These cymbals are compressed by strong muscles, so they buckle one way then the other as many as 4000 times a minute. It's almost like Rolf Harris' wobble board, but just the size of a sequin. That evening I found a cicada on the ground which had had its abdomen eaten out by some kind of parasite. I surmised a parasitic wasp had laid an egg on the cicada's abdomen. As I examined it under the magnifying glass, I noticed in horror the cicada's head start to move and its legs to convulse, clearly in agony. 'Holy crap' I stammered; 'it's still alive'! I was just poised to drop it underfoot and put it out of its misery, when a repulsive mucus-covered head broke through between the cicada's eyes, and the maggot that had eaten it alive from the inside out, wriggled out of its head and dropped to the floor as I watched in sickening close up. Sometimes nature is so much more macabre than a horror movie could ever be.

Axolotls: the little amphibian that never grew up

by Trevor Smith

This is a species that you either love or hate. They are considered by some to be super ugly critters, but I have a soft spot for these totally aquatic salamanders so I will tell you a little about them.

Originally from only two lakes in Mexico, Lake Xochimilco and Chalco, in the wild this species, **Ambystoma mexicanum,** is seriously under threat, as the lakes are in trouble. They are designated a CITES B endangered species. However, there are plenty of them in culture across the world , all of which it is said came from just 6 specimens originally collected by a French research scientist who bred them. Axolotls where actually part of the Aztecs' diet and apparently taste like eel, not that I have or will try them, far too pretty to eat! But they do make great pets.

The wild colour form is dark grey with almost black gills but now bred in several colour forms, black, white, piebald, yellow and albino to name the commonest.

These are actually a neotenic mole salamander, Neoteny, which is also sometimes called juvenilisation, is the retention, by adults, of traits previously seen only in juveniles. This is a little misleading with axolotls and probably comes from the confusion with another species of related salamander from the same regions namely the tiger salamander (*Ambystoma tigrinum*) which does look like the axolotl whilst in the larval stage and can, in fact, remain as the juvenile form for some time if poor conditions prevail during development.

However, tiger salamanders will metamorphose into adults eventually but axolotls never naturally become adult and breed in the larval stage.

Axolotls are quite large and grow to 300mm and can live to 25years so a nice impressive long term pet.

They have a large head, stout limbs and well developed gills; the tail is relatively short and laterally compressed. They have large dark

eyes with pink eyes in the albino form; this form also has red gills.

Totally aquatic in habit, they never leave the water. Fertilisation takes place in the water with the male dancing in front of the female before depositing his sperm capsule which the female swims over and collects. The spawn is placed on to aquatic plants and up to 1000 eggs can be laid by a single large well fed female, although I find 300 – 400 to be more the normal amount. Tadpoles hatch after about 20 days that look like miniature adults; these will reach sexual maturity in one to two years.

They feed on a whole range of soft bodied food including earthworms and slugs, but they will tackle anything that gets close to them that they can grab, including fish and do very well on floating or sinking pellet food for fish.

In captivity

This is a splendid species to keep and thrives in captivity, with records of up to 25 years, which is impressive for any amphibian. Although totally aquatic these long lived captives will soon acclimatise and learn to react to your presence. Once settled they will quite happily feed from your fingers, coming to the surface at feeding time. It's like keeping little crocodilians as they raise up to grab their food, and, oh don't worry if they grab your finger they will do no damage.

Cage set up

These are quite an sedentary species and you will require a reasonable sized aquarium to keep them in. I

ADULT BABIES

Axolotls exhibit a property called neoteny, meaning that they reach sexual maturity without undergoing metamorphosis. Many species within the axolotl's genus are either entirely neotenic or have neotenic populations.

However neoteny can also be found in other amphibians, including at least one British species of newt. In the CFZ archives are records of a neotenous smooth newt found in Cornwall in the mid 1990s.

In the axolotl, metamorphic failure is caused by a lack of thyroid stimulating hormone, which is used to induce the thyroid to produce thyroxine in transforming salamanders.

Unlike some other neotenic salamanders (Sirens and mudpuppies), axolotls can be induced to metamorphose by an injection of iodine (used in the production of thyroid hormones) or by shots of thyroxine hormone.

use a 48" long x 12" wide x 14" high tank to keep a trio in. To avoid having to do complete tank clean outs I add an under gravel filter. Aquatic plants set the whole thing off and an added piece of bog wood will add to the whole effect, just make sure you don't have any sharp edges. On that point, another remarkable thing with these salamanders is the fact that they have noteworthy regeneration capabilities and if they lose a gill or a whole limb they grow it back in quite a short time with no obvious stress. Heating is unnecessary as a room temperature aquarium is fine with water temperatures of around 60°F in summer and 45°F in winter being fine.

Feeding is easy as they will take earthworms, wax moth caterpillars, crickets, slugs and blackworms as well as most soft bodied or small bugs you can find under logs in your garden as well as strips of lean meat or fish (although these can really mess up the water and cause you to do a partial water change) young tadpoles will require smaller food and are best started off on daphnia or chopped bloodworm. They have good appetites and will need regular feeding to keep them healthy and for you to have a chance to breed them.

The trigger for breeding in this species is quite an unusual one. In their native lakes the first signs of the approaching spring is actually a drop in the water temperature, signalling the melting of the ice in rivers feeding the lakes. So in captivity well fed adults can have lumps of ice added to the tank in the spring and normally spawning takes place a couple of weeks later. The spawn is best removed on the plants it is laid on and placed in a well aerated tank with around 2" of water. The young will need to be separated as they grow, and during the first few months are highly cannibalistic, however once they are at around 3" they lose this tendency and settle down.

So if you are into keeping aquatics and want a long term pet that's easy to care for and a bit different from tropical fish in a tank, I would highly recommend you get to know a little group of these fascinating amphibians.

NOT QUITE

JAWS

Back during my tenure as Deputy Editor of *Tropical World* David Marshall submitted this article to my late boss. However, for reasons best known to him, he decided not to publish. This was a decision to which I took great exception, because I had greatly enjoyed reading it. So, it is particularly pleasurable to be able to publish an updated version five years later.. (and to finally use the stupid artwork that I planned so many years ago)

David Marshall

Background

Found in the freshwaters of Asia and Africa are a number of primarily vegetarian members of the Cyprinidae family (Carps and Minnows) whose body contours resemble, to various degrees, those of the marine sharks, thus they have also gained the tag of 'shark' within their common names.

Until recent times Ichthyologists tended to place the majority of these particular 'sharks' within the genus *Labeo,* but ongoing revisions are now, albeit at a trickle, starting to split them into more discernable groupings. To see this in action we only have to look at a very old aquarium favourite that goes by the common name of the red-tailed black shark. Asked to verify the scientific name of this lovely fish we long-time aquarists immediately plump for *Labeo bicolor* only to find that the rather longer scientific name of *Epalzeorhynchas bicolor* now applies.

As these 'sharks' vary greatly in both their overall body size and natural habitats they offer the aquarist a wonderful opportunity to mix and match their kind in order to suit a wide variety of tropical set-ups. However, as I can verify through experience, never jump into such a purchase without first thoroughly researching the specie concerned, as their temperament, attitude towards their own kind, overall body size and specific aquarium requirements may turn out to be very different from what you expected. For these reasons I specifically chose to cover the seven species featured in this article.

Epalzeorhynchas frenatus

The ruby-finned or rainbow shark, which grows to a size of 12cm, is the most readily available 'shark' in the aquarium hobby. Originating from Thailand these fish are available in both natural and albino form. These fish should always be kept in small groups with aggression thus spread out among the group members and not towards an individual. An aquarium of 90x30x30cm will house such a group in the company of *Trichogaster* gouramis and large loricarins. They prefer slightly alkaline, rather hard water conditions but are not 'fussy' as to an exact tropical temperature. Easily fed upon commercial aquarium flaked and granular food.

Within a group what appears to be spawning behaviour will be seen at frequent intervals. However, spawning in aquaria is very rarely known. Many of the fish seen for sale have been 'developed' through hormone stimulation techniques used in large scale breeding establishments in the Far East, Florida and Eastern Europe.

Labeo chrysophekadion

This fish is actually a reversal of the ichthyologic revisions as it has moved out of the genus *Morulius* (leaving *M. calbasu*, the sagarmatha shark, as the lone member of this genus) and back into *Labeo*.

The common names of black shark and black sharkminnow originate from the black juvenile colour stage. By the time the 90cm adult size (tail length) is reached, (usually smaller in captivity) the body colour has faded to plain grey and is highlighted by one iridescent spot upon each scale. At the present time golden and xanthic variants, both of which were first developed by commercial breeding establishments in Florida, are also available but please note that the hormonal stimulation breeding technique used to produce these

fish lessens their lifespan in comparison with the 'natural' colour form.

Found in rivers, streams and canals the black shark has a wide natural distribution, which takes in the Malay Peninsular, Java, Borneo, Sumatra, the Mekong basin and the Chao Phraya basin. Mature adults' time spawning runs to coincide with the arrival of the rainy season and once the fertilised eggs hatch, the fry dine on an abundance of micro-organisms found amongst flooded grassland.

Before taking on a black shark, with the requirement of an extremely large aquarium to house a single adult, please think very carefully, because although they can become very tame towards their keeper (and be trained to play with tiny balls of tinfoil) and are long-lived, they have their downside. The first of these is that they can quickly decimate any décor, so silicone seal large pieces of coal, large plastic plants and synthetic bogwood to the glass base. Secondly, I learned the hard way about just how the temperament of a black shark works! In the early 1980s I set-up a Rift Valley community stocking this tank with large *Cynotilapia* and *Nimbochromis* species. Having read that a black shark would fit into such a system a beautiful 5cm long youngster was purchased. It took the shark a fortnight to dominate and have the cichlids corralled into the four corners of the aquarium. Every time a cichlid moved it was relentlessly pursued and the bluer the body colour the more harassment was dished out. As soon as a spare aquarium matured, into this the shark had to go! The cichlids soon recovered.

The black shark requires a high temperature, around 27 C, and a pH of between 6.5 and 7.5. From experience I recommend keeping a single black shark either on its own or in the company of large catfish that are well established in the aquarium before the shark arrives.

Labeo cyclorhynchus

To find further information upon this specie you may have more joy searching under the junior synonym name of *Labeo variegatus*. Because of the dark grey and yellow speckled appearance, supplemented by reddish fins of juveniles, this fish has the common names of harlequin shark and harlequin sharkminnow.

This African fish comes to us, wild caught, from the Middle Congo, the Lower Congo and the Ogowe river system of Gabon. During the 1950s the harlequin shark became the first African *Labeo* to become available to aquarists. Now, 50 years on, we are in a position where hobbyists and scientists alike, have built-up more knowledge of its care in aquaria than is known of the wild existence and spawning habits of this specie.

It did not take long for the aquarists who pioneered the keeping of this shark, to report that here also was a fish that took badly to the presence of its own kind, and which was capable of showing the same level of a gross intolerance towards other fish species as we talked of with the black shark. So why has the harlequin shark remained a popular aquarium fish for so long? The answer lies in the fact that these are probably the most beautiful of the freshwater 'sharks' available in the aquarium hobby. Look closely at a group of juveniles and you will see how greatly both the body pattern and depth of colour can vary between the individual group members. Even at the fully-grown size of 30cm they are a sight to behold and by now the body has developed into a velvet-grey colour with a red spot on each scale.

Although the harlequin shark is not as deep bodied as its black shark cousin, the aquarium size and conditions remain the same. I have heard of aquarists who have had some success in keeping these fish as small groups in large aquaria, in which a multitude of caves and hiding places were placed, but I do not advise taking this risk so stay with a single Harlequin.

Labeo rohita

Labeo rohita, the rohu shark, can grow to a size of 200cm (tail length) and have a body weight of 45kg that makes it an important aquaculture species. To properly house one of these fully grown 'giants' would call for a custom built aquarium, complete with high turnover filtration.

The rohu I cared for was well settled at a pH of 7 and its aquarium was kept at a constant 27°C (although it is known for this specie to live at a temperature as low as 14°C). It ate large quantities of vegetable-based flaked food, cichlid pellets and adored pieces of Thai crabstick. It never attempted to disrupt the décor and lived in almost perfect harmony with a group of large silver dollar, a *Distichodus affinis*, a pair of large tinfoil barb and several *Synodontis*.

What I remember most about this particular fish was that when it became 'spooked' by any of its companions it immediately dived into the gravel and literally buried itself. Like a number of large cyprinid species, it had an almost inbuilt instinct that cats were bad news and every time the sound or vibration of paws was detected on the fish house roof, the rohu would be very nervy for several hours. I never dare lift the condensation shield at this time as these fish are powerful leapers!

Labeo boga

The velvet or red gilled shark is a remarkable fish. Capable of reaching a size of 30cm these fish are found in the rivers of the Gangetic Provinces of India.

The velvet shark will not disturb the décor and is not as aggressive as some of the other sharks we have looked at, but does display the nasty habit of 'fin nipping'.

The majority of specimens, which arrive into the aquarium hobby, do so as fingerlings from fish farming operations.

Leptobarbus hoevenii

The cigar shark, which you may also know under the names of Thai carp and hoevenii barb, is probably the tropical cyprinid that is most purchased by aquarists who do not realise that the cute 4cm long little barb they have brought home can reach a total size of 50cm! There is little wonder that juvenile cigars draw the aquarist into a purchase as they have a lovely metallic matt body, red in the majority of their fins and what appears to be a double-lined lateral of thick black topped with a bright gold-green. As the fish mature so their colours dull.

In the wild, the cigar shark is found in Borneo, Thailand and Sumatra. Many of the cigar sharks we find in aquaria are the product of aquaculture projects. Again, we set-up a tank as per the black shark, but have little to worry about disturbance of the décor as these fish will stay in the surface and mid-water areas. We aim for a lower water temperature of 23°C. Good filtration is a must. You will have no problems with feeding as commercial cichlid pellets, vegetable-based flake food and occasional feedings of wheat germ koi pellets are all accepted with great gusto.

However, please note that cigar sharks are brilliant escape artists which, given the opportunity to do so, can leap out of and be found some distance away from their aquaria. They are also very prone to a specific form of protozoan 'whitespot', the same 'strain' that also causes great problems with *Microsynodontis* species and *Synodontis brichardi*, which mats the body so violently that it is impossible to cure, so never mix with fish which are known to be prone to protozoan problems.

Here we have a freshwater 'shark' that will live with its own kind however. An ornate birchir and a variety of large catfish formed the company for the cigar shark which came into my care and this community worked very well. Never keep as companions any fish species, which may have very quick or sneaky movements, as this activity can badly 'spook' our subject specie and although a very rare occurrence, it has been known for a 'spooked' cigar shark to crash through aquarium glass.

Balantiocheilas melanopterus

No article on 'shark tag cyprinids' would be complete without a mention of the silver or Bala shark. This 'frightened of it's own shadow' cyprinid comes to us from Thailand, Sumatra and Borneo. This powerful jumper, with the ability to attain a body length of 40cm, is found in several natural variant colour forms based on a bright silver body with white fins that are edged in black. Although I am not advising you to 'try this at home' the people of Asia remove the delicate silver scales and use these in the production of 'mock pearls'.

Well they may not be 'jaws' but the freshwater cyprinids with a 'shark tag' are an interesting bunch of freshwater fish and I hope that this article will inspire you into further study of the species covered.

The Jade Mantis

Hierodula salomonis Werner, 1930

by Graham Smith

This species is new to the hobby and is still yet to be commonly seen. It is originally from the mountainous tropical forests regions of Guadalcanal - which is the largest Island of the Solomon Islands - forming part of the third largest archipelago in the South Pacific, with 992 islands and a total area of 28,450 square kilometres.

This species has the most stunning range of colours from pink through to pale yellow, and into the greens, and all with a depth that can only be likened to that of jade, hence the common name. The body shape is a typical mantis with no foliate extensions or other added features, as living amongst the tropical vegetation its colour is all that it needs to blend in and vanish when not moving.

This is a medium sized mantis, with males and females both reaching around 75mm when adult, although

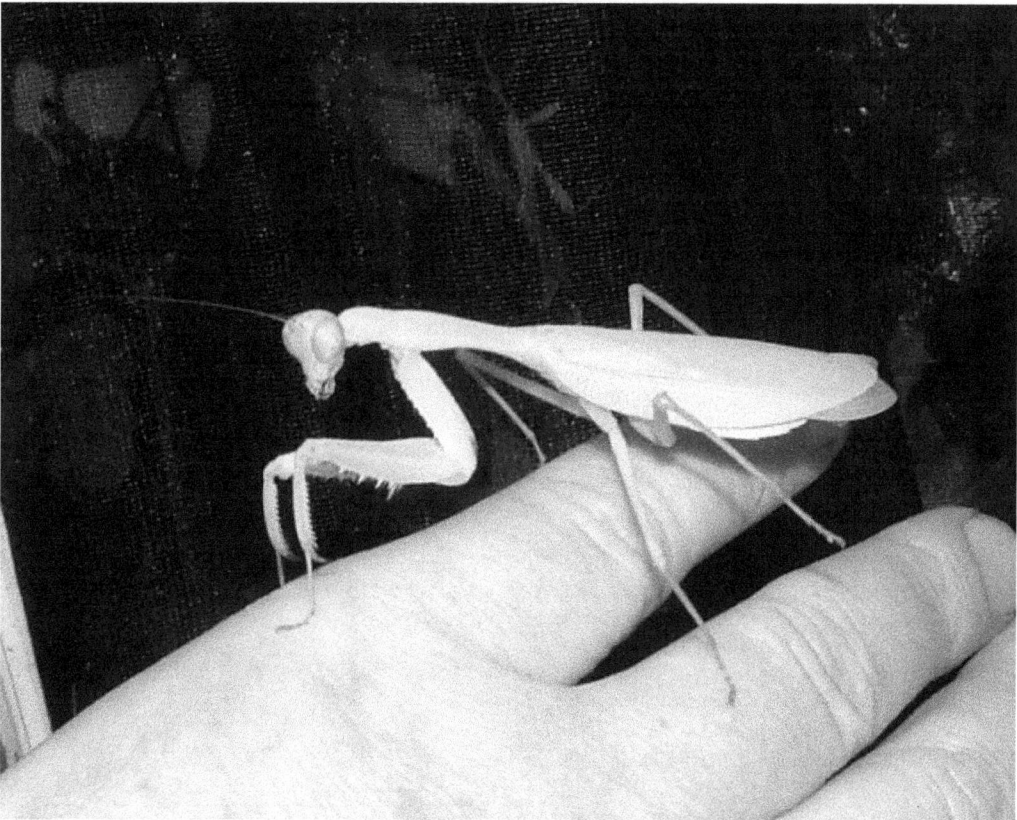

as per the normal with most mantids the males are of a slimmer, lighter build. This lighter build allows the males to find the females by flying to them, which the females - although fully winged - cannot do once they are at breeding weight.

At the time of writing this we are the only people with this species in culture, even after getting it to other breeders, and all of our culture comes originally from a single oothecae collected in the wild, but we hope to get it stable enough to make it available to everyone.

Keeping Jade Mantids in captivity

Coming from the tropical forest, I like to use a slightly moist substrate to give some humidity for this species, and provide a temperature of around 25°C-30°C. Being quite a shy forest species, I also like to add plastic, silk or real leaves into the cage otherwise if you approach the cage these mantids can panic and try to escape, causing them to refuse food. With the leaves, you can observe the way they work in the wild. As you approach they will pull themselves flat to the leaf and it's a pretty good trick, as they really do fade from view. I give the mantids a light spray of water once every second day just to wet the leaves as this gives them a chance to drink, and also keeps the humidity up.

The cage can be quite a simple affair and it is best to use something like a glass tank, as this species can be a little nervous and like to feel that they are hidden from view. In fact this species can benefit enormously by having living plants and bog wood as part of their cage furniture.

Although stunningly coloured when out of their natural environment they are in fact quite a well camouflaged species of mantis which allows this species the chance to not only survive as an ambush predator, but also as an active hunter stalking its intended prey.   In captivity (as in the wild) they will happily feed on fruit flies, midges and mosquitoes as nymphs and, as they grow, so does their food size, eventually taking Katydids, grasshoppers and flying prey up to

THE SOLOMON ISLANDS

The Solomon Islands is a country in Melanesia, east of Papua New Guinea, consisting of nearly one thousand islands. Together they cover a land mass of 28,400 square kilometres (10,965 sq mi). The capital is Honiara, located on the island of Guadalcanal.

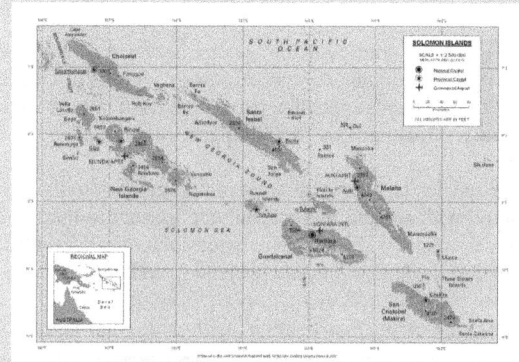

The Solomon Islands are believed to have been inhabited by Melanesian people for thousands of years. The United Kingdom established a protectorate over the Solomon Islands in the 1890s, mostly for humanitarian reasons, in order to put an end to a type of semi-slavery called `Blackbirding` which was a trade in captured natives forced to work in Australia or Fiji. Some of the most bitter fighting of World War II occurred in the Solomon Islands campaign of 1942–45, including the Battle of Guadalcanal. Self-government was achieved in 1976 and independence two years later. The country is a Commonwealth realm, and Queen Elizabeth II is head of state.

the size of a locust, but they will also happily tackle crickets and bluebottle flies and although these are not a natural prey item they work well.

Breeding presents no problems, providing the females are at least 5 weeks old (that is five weeks from the final moult into adult) and VERY well fed. The male can be introduced into the female's cage and left along with plenty of live food. Once mated the male should be removed into a separate container for his safety and the female should be fed all she can eat. Females will lay up to 4 oothecae over three months. These are white at first turning to pale beige then on to beige brown, and are around 40mm long and 30mm wide and will hatch with little problems in 4 – 5 weeks if kept under the same conditions as the adults, producing 50 – 150 young.

The oothecae are quite robust and are best removed from the cages a day or so after they have been laid. It is best to remove them on the stick they have been laid on and placed in a small clear cup or container with some Bluetac, with some slightly moist tissue on the floor, some fine mesh on the top secured with a band and sit back and wait. The nymphs are almost white when they hatch and turn brown a few hours later. Cannibalism is quite the normal for this species, so it is best to house the nymphs separately to avoid excess loss. These can either be housed in a similar set up to the adults or kept in simple containers, like a small clear cup, with a little tissue on one side a small stick and a thin layer of substrate. Cover this with netting secured with a band.

They moult and grow at quite a fast rate and take on some of their superb colourings from around the fourth moult and reach a gorgeous mixture of colours at sub adult, which is mainly hidden when they reach adult and the wings cover their backs. However, at this stage, they take on the true jade look.

This is a new species to culture, but I have no doubt that providing we can add fresh blood lines into our stock to prevent genetic burn out I am sure it will become a much prized species in the hobby, and will always have a place with me.

Nymph (Rob Byatt)

Snakes with fins

by Max Blake

The order Perciformes contains a total of 9,300 (and rising) species and is thus the largest vertebrate order. This short article will attempt to do justice to my personal favourite family within that great order, Channidae; also known as the Snakeheads. If you read the grungier pages of tabloids or the US media, you may have seen these fish called "devilfish" or "Asian Terrors" after they were found to be breeding and overwintering in US waters. The media loves a good story, and quickly snakeheads were hyped up to be a terrifying enemy capable of walking over land, surviving three days without water, killing humans and out-competing native fish. Two of those points are true, one is slightly true, and one is complete rubbish; hopefully by the end of the article you will be able to work out which is which!

Dwarf snakeheads are from Southern Asia (with an allied genus to the more common *Channa*, *Parachanna*, from Africa) and make superb pet fish. Hang on, *pet* fish? Surely that word means that snakeheads must be intelligent fish with individual personalities? Quite correct. "Dwarf snakeheads" is a catch all term for the *Channa gachua* species group, a name for snakehead species with alternating dark and light bands on the pectoral fins. They are also characterised by being mouthbrooders in the same vein as some cichlids, except that it is the male snakehead who holds the young. This is their prime method for sexing: adult males have more bulging cheeks compared to females. None of this group grows larger than 12", and most max out at about 7". The smallest species of the group, *Channa* sp. "Assam", only grows to a maximum of 4". It is being hailed as a new species, although I have to say it does look very similar to *C. bleheri* and

Channa ornatipinnis

Channna orientalis

may end up being a sub-species.

Due to their chilled out nature, the largest dwarves can be kept in a 4ft long tank, and the smaller ones will be happy in a 2ft. They have a low dissolved oxygen intake, and instead rely much more on their labyrinth organ to breath atmospheric air. Although this structure has the same name and purpose as the anabantoid's structure, it has evolved independently and the two groups are classified in separate sub-orders. This, along with their long but flat body, helps them move over damp ground, but only when the humidity is high (monsoon season). The three days idea the media popped up with is total rubbish.

Feeding is down to personal preference. Sorry, not yours, the fishes. Some will take only livefood, some frozen, some pelleted and, very rarely, flake foods. Any meaty food will do, and even for those that eat pellets some meat (blood worm or prawn) is a good idea. Adults usually need feeding every other day. They are very greedy fish, and will take as much as they can and get very fat; something to be avoided. Their large mouths let them take comparatively large fish while their powerful bodies give them huge thrust for catching prey. They can jump out of the water, and one of mine has learnt to jump for his supper when it's stuck to the tank roof. A rough guide to tankmates is anything over two thirds the snakehead's size and not slender. Large shoaling fish and other non-aggressive oddballs work well. Dwarf snakeheads can be kept in groups, but this requires a large tank and a lot of fish. *C. bleheri* is the species happiest to be kept in a group. Generally though, one snakehead per tank is the rule to follow.

Snakeheads do have a reputation for being brutish fish. This is slightly true. The large species (2ft long and over) can be aggressive, especially when they are defending their young from predators. Adult *C. micropeltes* have been known to kill people swimming too close to their nest inadvertently. If you have ever seen a big *Channa*'s teeth, you won't doubt this! Dwarf snakeheads are generally non-aggressive, but there is always that odd one who follows his own rules.

As a brief interlude, I once watched a documentary about the problems America was having with foreign

Channna orientalis

Channa pulchra

Channa pulchra

snakeheads. The name of the program I had best not say (although, if you were to, say, "fish" and "zilla" as one word, you may find some information about it), but it was horrific! A case of terrible over-hyped reporting if there ever was one. Americans generally think it is fine to use live fish to feed to predatory fish within the confines of aquaria, but this is not the place to discuss such practises. Anyway, the director had found a clip of three sub-adult *C. micropeltes* (the giant species) eating goldfish. This was stuck this into the program, and the narrator was told to say *"Snakeheads are vicious and devour their tankmates."*. This was so daft I remember I burst out laughing! Another bit interviews a local fisheries manager who captured four adult snakeheads from the river and kept them in a large vat with some steel mesh covering the top. By the morning two of the snakeheads had left the tank and were moving around on the floor. This, apparently, was evidence of how they can easily escape any confinement. Now just think this through for a second. Two snakeheads were left in the tank. The fish were adults. Snakeheads hold territories. What happened? Two of them paired up, and attacked the others such that they jumped out of the tank to escape their wrath. Simple. Anyway...

Most species of dwarves are tropical and do well in tanks with lots of decoration and a slow flow. They are not really fussy about water quality, and slightly hard and alkaline water will be fine. Common species include *C. gachua*, *C. bleheri* and *C. orientalis*. Recently there was an attempt to ban all snakeheads from entering the UK after the problems in the US, but this idea was dropped begrudgingly when someone told DEFRA that most snakeheads are stenothermic, and cannot cope with UK temperatures. *C. argus* is an exception, and can go down to about 17°C. There was an oafish hoax a couple of years ago involving a UK angler and a dead Argus snakehead, but this was so daft it was laughable.

New species to the market can be very expensive, but they are mostly sub-tropical (20-22°C) and need cool water with a high flow of clean water in a rocky environment. Such species include *C. ornatipinnis* and *C. pluchra* (illustrated). If you do see a snakehead unlabeled in a tank, chances are if it has the banded pectoral fins, it will be a tropical dwarf species. If it doesn't have the bands, don't touch it! Chances are it will turn out to be one of the huge species, which although great fun to keep, need a really long term commitment.

TREVOR'S TAILS

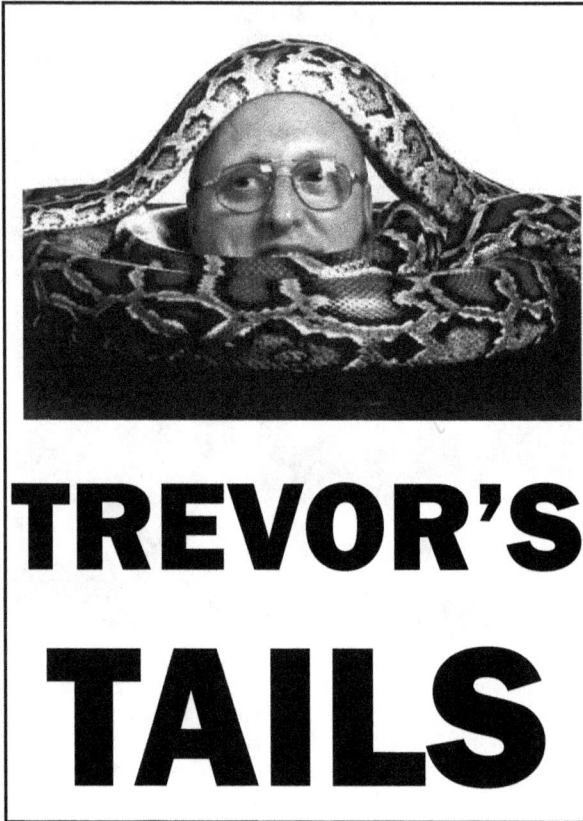

Hi all, and a belated Happy New Year to all of our readers.

I hope you all have had time to do a little something for our own wildlife during this winter, which I think I am right in saying is the worst for over a decade. So if you haven't yet done anything, at least fill the garden with bird food and water and if you haven't got a garden, just put it on your roof or on your parked car or anywhere they might find it.

Filming is still as crazy as ever with me and, as always, I am asked for species out of season. This week it was caterpillars and *"they must be an English species"*. Sometimes makes me wonder if the filming companies believe in seasons! Also I have had the chance to go along and take some of our menagerie to kids over the Christmas break. I always get a great kick out of this - just the look on their faces when they get to hold a big snake or meet some of the giant cockroaches or any of the other animals.

You know it's something they will remember and talk about for a long time and instead of thinking they are 'horrid' come to find them interesting and hopefully will have the interest for life.

Well other than that it's been an interesting start to the year with a house move and subsequent moving of my menagerie. Now I try not to keep many things at home, but a house isn't a home without some exotics around, or at least it isn't to me.

One of the benefits of being am amateur naturalist is that we remain fairly credit crunch proof. I say this as everyone else seems to be saying *"its the end of the world as we know it"*, and us naturalists are simply saying spring is on its way so the birds will be nesting soon and when is the first animal show on? So, let's all relax and remember the things in life that are truly important, which is life itself rather than money and possessions.

Having just re-read what I have written I think it shows that I am just an old hippy, and to be honest I'm quite proud of that. So, Peace and Love to all and let's have a superb and enthusiastic 2009.

LUCY'S LIFE

Lucy is a young lady of sixteen, who appreciates the important things in life; she breeds mantids, has a bevy of escaped hamsters under her bath, and has a little brother who is obsessed with snails. In short, the perfect columnist for this magazine…

Well, even though it is february I would like to wish you all a Happy New Year! Let's hope it's a good one!

A lot has happened since the last edition of *Exotic Pets*; for one thing I went to my first show! It was the Doncaster IHS event, and although I promised I would be sebsible, I failed miserably and spent *way* too much. However I came away with two beautiful leopard geckos; Dexter and Brian, and a lovely corn snake, Snakey.

As many of you know I had a hamster escapee – Eddy, (who sadly passed away the other month). Well, Snakey decided to go on a little tour of my room and was missing for nearly a day, how ironic was it when I found him buried in all of Eddy's old things!

I then got another 3 snakes! Charcoal, Kikoho and Dave.

Unfortunately my Tarantula collection has taken a turn for the worst! I recently found out I have a very bad allergy to the urticating hairs, so I had to sell most of the collection. I have kept Bubbles (She was my first T), Ginge, Smarties and Freddie because they were all given to me as gifts.

I traded most of my T's for millipedes as I guessed they were the only other invert I hadn't kept on the "Bugnation checklist" I'm finding them to very interesting but very slow growers. I'm happy with my current collection so won't be expanding any time soon (she says).

While on the subject of inverts, I recently got some flower beetle grubs, so it'll be interesting to see how I get on with them! Looking forward to spring/summer as hopefully my *H. membranacea* mantids will be adult, and I'll be able to start breeding them again!

Another new animal I have added to the collection is mice! I have three females, which one gave birth not long ago, to seven lovely babies! And yes, I've had one escapee already! (So far) I don't know how all my animals manage to escape, but it makes them more at home I suppose.

Anyway that's all from me, hope you enjoyed this episode, and hope the february chills aren't too cold!

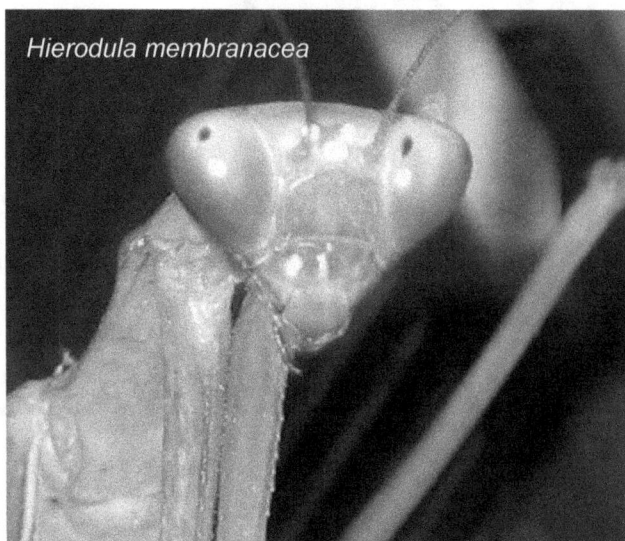

Hierodula membranacea

Bugfest SW

The Carnival of Monsters

Come and meet amazing creatures from Earth and special guests from outer space...

www.bugfest.co.uk

Cyberman © BBC

Saturday February 21st 2009

10.30am – 3.30pm

Stanchester Community School

Stoke sub Hamdon TA14 6UG

Admission: Adults: £5   Children: £3

Family pass: (2 adults 2 children) £14

10% from admission sales goes to The Somerset Children's Hospice

Charity number: 1007710 Registered in England No: 2820879

The Bengal Spiny Stick insect
Spinohirasea bengalensis (Brunner, 1907)
by Janice Holt

S*pinohirasea bengalensis* is quite a new species to the hobby in the UK but I am sure it will soon become a firm favourite with keepers.

This mostly nocturnal species comes originally from the central province of Ha Tinh in Vietnam, a fairly populated, but quite tropical, farming region, but as the name suggest it also comes from a fairly wide area across tropical Asia. It is a highly cryptic species with many spines and projections that in the wild help it to vanish from view and in captivity stand out like the beautiful gem that it is.

Overview

These bulky, if somewhat small, species has females that are larger than males (reaching around 70mm, with males at around 50mm). It is a little different in appearance from the standard stick types, and have an overall green brown ground colour with a multitude of spines. The males are slightly less heavily built than

the females and have fever spines; both sexes have hints of pink on the flanks and quite a red belly, especially the males. Neither sex is winged. This species reproduces by sexual reproduction, with the male staying on the female's back for days on end. During its lifetime, a female may lay several dozen fairly large eggs all of which she lays by simply dropping them on to the soil.

Captive maintenance and diet

As with most stick insects they require a simply cage that is at least 300mm high to allow adequate space for hanging down to moult. Other than this you can use a sweet jar as the simplest cage, through to an elaborate glass tank set up with living plants etc., but do consider the fact that the more elaborate the cage the more of a problem sorting eggs and cleaning becomes.

I keep my insects in sweet jars with mesh lids. On the floor, I put a thin layer of slightly moist kitchen paper; this allows for easy cleaning out and allows me to find the eggs. Food is placed in a jar filled with wet vermiculite to help keep it fresh, this species will eat quite a range of foods including bramble, ivy, rose, hawthorn to name a few, but I have found that if in doubt just offer a new food with known foods and if they like it they will eat it!

I tend to keep lots of jars set up in the same way with each species and start up a new one for nymphs as they hatch and move on as they grow. This way you can have twenty or thirty young nymphs or up to 10 adults in a jar. This also makes feeding a little easier as I empty each jar and check the individuals at each feed splitting them up as necessary and putting them into freshly set up jars. Temperatures should range between 22-28°C.

Breeding

This is another very easy species to keep, rear and breed readily with the only problem being how many are produced. However, there are normally plenty of people that want to keep this species so finding homes for them is rarely a problem.

Eggs are dropped on to the substrate from where they should be removed and placed in containers with slightly damp substrate where they will hatch with great success.

Eggs typically take about four or five months to hatch. Nymphs hatch at approx. 1.5cm long and are quite chunky for a fresh nymph. Although they should be fine in with the adults, I like to house mine separately so I can monitor them better. They are typically voracious and grow surprisingly quickly, reaching adult in about 4 months at 25°C and are mostly trouble free. The only thing to avoid is over crowding as this can lead to limb loss with individuals eating limbs accidentally whilst feeding or through disturbances whilst moulting.

In Conclusion

Spinohirasea bengalensis is a great insect being attractive, hardy and a prolific phasmid that, once it is well established in captivity, it will be a must for the first time keepers, as it has all of the right ingredients, very pretty nymphs that are easy to keep and breed, with needs that are easily met.

I highly recommend giving this species a try, but then I would wouldn't I?

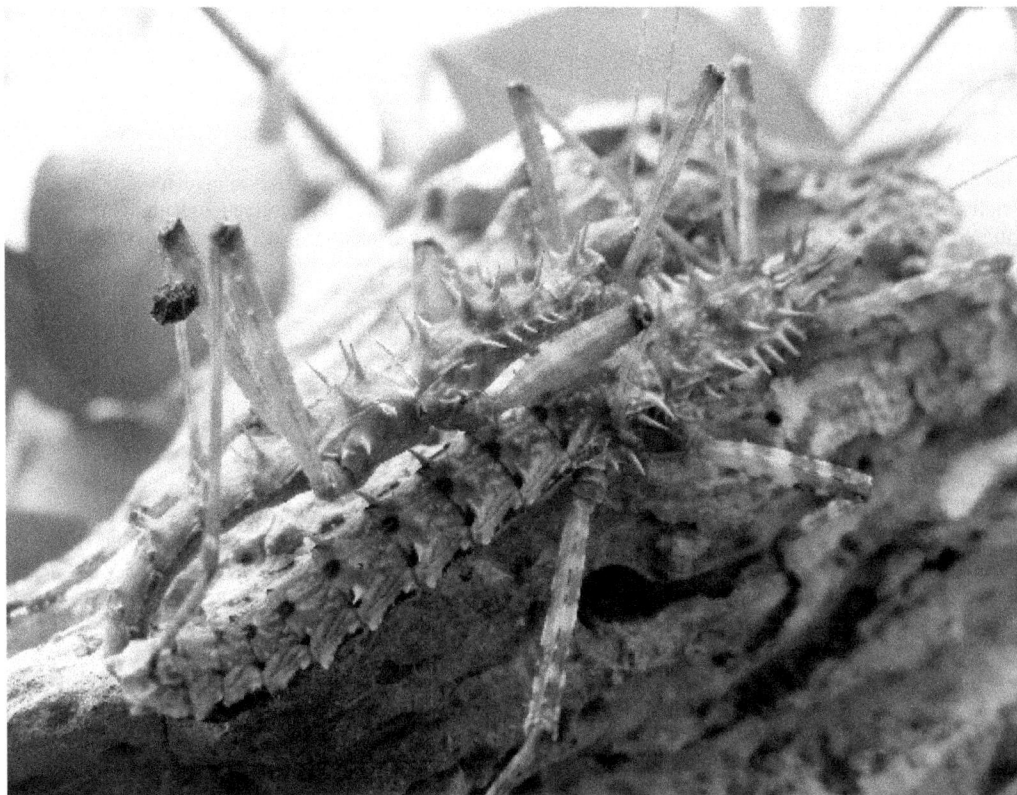

Your First Snake
by Richard Freeman

S nakes can make interesting and rewarding pets. They are beautiful and, with handling can become quite tame. In this article I will look at how a first time snake keeper can go about choosing and caring for their pet.

THINGS TO CONSIDER

You will need to look at a number of factors before buying your snake.

- How large a vivarium can you have? Do you have enough space in your home? Snakes can differ in size vastly from the 4 inch Lesser Antillean Thread snake, *Leptotyphlops carlae* to the green or giant anaconda, *Eunectes murinus* at 30 feet (and possibly considerably more).A beginner would choose a fairly small snake, but even the species listed here can range from a little over a foot to around six feet in length.

- The Cost. Even an easily kept snake will need equipment such as heat pads, lights and thermostats can you afford these?

- Feeding. All snakes are carnivores and need to eat pinkies, mice or other animal food items. Can you get these? When keeping snakes you cannot be squeamish, there are no vegetarian snakes!

- Always think before you buy an animal and never buy on impulse. There are a number of good books and websites that give information on snake care. Read up on the subject before you visit the pet shop.

- Always examine your snake carefully before buying it. Look for things like mites or ticks.

- Examine how clean the shop is and how in what conditions the animal had been kept. Ask about how well it has been feeding and how often. Never buy a snake that looks boney. Also ask where the snake came from. It is best to buy captive bred animals that wild-caught ones.

Now, lets look at some good snakes for beginners.

THE COMMON GARTER SNAKE *Thamnophis sirtalis*

Found throughout most of North America, this is an idea snake for beginners. It is small, generally feds well and is easily tamable. You will need a 15-20 gallon vivarium for one snake. Double the size for a pair. Aspen wood chips, corn bark, or a commercial reptile substrate can be used to line the floor of the vivarium. A water dish in which the snakes can submerge is essential. You can decorate the vivarium with rocks, ferns and mosses. Due to their small size, garter snakes do not tend to crush live plants when they crawl over them. However, you must make sure that your vivarium has a secure top. Snakes are the greatest escape artists in nature!

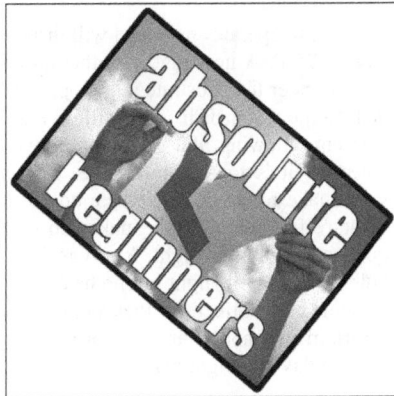

This is a temperate snake and will thrive at a heat of around 75°F. A heat mat with thermostat should be placed under the substrate at the opposite end of the tank to the water dish. This will give a temperature gradient allowing the snake to move around to wherever it is most comfortable. Unless you want your snake to start to hibernate do not let the temperature fall below 70°F. Snakes do not usually require an ultraviolet light. You can buy specialised bulbs that give off light but no heat. Thus heat from a light bulb affecting the ambient heat of the vivarium will not be an issue. Remember to have a marked day and night period.

Garter snakes can be fed on a mix of live earthworms, defrosted pinkie mice, guppies, and small goldfish. I find feeding with tongs or tweezers is best. Feed once per week. Garter snakes tame down quickly, but one that is not used to being handled may eject a foul smelling musk from its cloaca. This is a defense mechanism used to put off predators.

HOGNOSED SNAKES (*Hetrodon sps*)

A number of species of hognosed snakes are found in the Americas and Madagascar. The best for the beginner are the eastern hognose, *Heterodon platirhinos* and the western hognose *Heterodon platirhinos*. These are small, stout snakes that grow from 15 to 24 inches as adults. They have a mild venom but it is not powerful enough to harm humans. In my experience, hognosed snakes are the most placid of all snakes and the least inclined to bite. They are usually quite tame and easy to handle,

They will hiss if they feel threatened. This hiss sounds very like the rattle of a rattlesnake. Taken with the animal's stout built and viper shaped head, this is a clear case of Batesian mimicry. This is were a harmless animal resembles a dangerous or venomous one in order to keep predators away.

The name hognose is derived from the upturned point on the animal's snout. This may act like a ship's prow and help as the snake crawls through the undergrowth.

The general set up for a hognose is like that of a garter snake. Aspen chips or a cypress mulch can be used as substrate or alternatively a commercial brand of reptile chipping. Hognoses burrow so give your snake some depth of substrate. Always provide

BATESIAN MIMICRY

Batesian mimicry - typified by the hognose snakes pretending to be far more venomous vipers - is a form of mimicry typified by a situation where a harmless species has evolved to imitate the warning signals of a harmful species directed at a common predator. It is named after the 19th Century English naturalist Henry Walter Bates, after his work in the rainforests of Brazil.

Batesian mimicry is the most commonly known and widely studied of mimicry complexes, such that the word mimicry is often treated as synonymous with Batesian mimicry.

Though visual mimicry more normal, acoustic mimicry is also known, and occurs in a variety of species. Predators may identify their prey by sound as well as sight, and mimics have evolved that play tricks on the hearing of those that would eat them.

One such case is the Burrowing Owl (*Athene cunicularia*), which nests in the ground. This species gives a hissing call that sounds much like a rattlesnake, which often spend the day underground.

Both the parent and young make such hissing vocalizations when threatened.

One predator of nestlings is the Douglas Ground Squirrel, which is duped by this auditory imitation

(Conant 1958)

Eastern Coral Snake (venomous)

Scarlet King Snake (non-venomous)

a water dish large enough for the snake to submerge..

The heating and lighting for a hognose are the same as above but the warmer end of the vivarium needs to be slightly hotter than for a garter snake, in the range of 82-85°F. Hognoses appreciate large rocks on which to rub themselves whilst shedding.

Hognoses in my experience are the least fussy eaters of all snakes, and I have never known a healthy hognose have any problems in feeding. Rat pinkies may be fed to small individuals whilst larger ones can be fed rat fuzzies or small mice. Feed once per week. Unlike other snakes, hognoses swallow their prey sideways in a manner most comical to watch.

THE CORN SNAKE *Elaphe guttata*

This species is one of a large croup of Colubrids known as rat snakes. It is probably the most widely kept of all the rat snakes. In the wild it can reach six feet but they seldom grow this large in captivity. Corn snakes are usually an attractive brick red colour but some can be gray. In captivity albinos are bred specifically for their whiteness.

Corn snakes can be a little more snappy than garters or hognoses but the usually calm down with handling.

A larger vivarium is required for these snake, approximately twice the size of the one quoted for the garter snake. The substrate, water and decor can be much the same as with the garter snake. Corns snakes excel in escape, even among snakes so keep that lid secure.

A heat mat and thermostat set up is the best way of heating a corn snake's vivarium but its heat requirements

A colubrid (from Latin *coluber*, snake) is a snake that is a member of the **Colubridae** family. It is a broad classification of snakes that includes well over half of all snake species on earth. Colubrid species are found on every continent, except Antarctica.

The Colubrids are certainly not a natural group, as many are more closely related to other groups, such as elapids, than to each other. This family has classically been a taxonomic dumping ground for snakes that don't fit anywhere else.

are slightly higher than either the garter snake of the hognose. The warmer end of the vivarium needs to be around 86-88°F whilst the cooler end 76-78°F. Corn snakes are happy with average room humidity (40%).

Adult corn snakes can be fed on mice or weaner rats. Most individuals will take between one and three mice at one sitting once per week. Feed with tongs as corns snakes can become excited at feeding time and can accidentally nip their owners whilst striking at food.

THE ROYAL PYTHON Python regius

If you are feeling more adventurous you may want to try keeping a constrictor. The most tractable and practical is the royal python. It is also know as the ball python due to the habit of wild specimens coiling up tightly when they feel threatened. The term 'royal 'python comes from the legend that queen Cleopatra wore these docile snakes as living jewellery. The Igbo people of Nigeria venerate this snake as symbol of the earth. They will burry dead specimens in tiny coffins.

We tend to think of pythons as gigantic and some of them are, reaching 33 feet or more, but the royal python is the third smallest python reaching a maximum of only five feet in length, and with an average length of four feet. Royal pythons are proportionately far stranger than a Colubrid of the same length. Being a constrictor it's coils are bulky and muscular. It is not large enough to harm a human but it can easily force the lid off a vivarium so extra security is needed in setting up a tank for a royal python.

A royal python vivarium should be at least four feet long with a similar height to allow the snake to climb. They appreciate sturdy branches to crawl across. They also like hiding places such as hollow logs. This species comes from the jungles of West and Central Africa so a substrate of aspen chippings, corn bark or reptile bark is needed. As rainforest animals they require some humidity. Humidity seems to be a factor in encouraging royal pythons to feed. Some keeper place tubs full of damp sphagnum moss into their vivariums to help keep the humidity up. A decent sized water bowl is also needed. As an African species they require tropical temperatures. The warmer side of their vivarium needs to be at 88-90°F with the cooler at

78-80°F. The ambient temperature can be dropped to 78-80°F at night. Royal pythons are nocturnal and do not require a UV lamp. They will, however, need a light source with a defined night and day. A caged ceramic heat lamp or under-substrate heat-mat can be used, provided both are regulated with a thermostat. Another way of heating is the 'reptile radiator'; a flat, square plate that screws into the top of the vivarium. As it is flat, there is nothing to coil around, so the animal will not burn itself. HabiStat are currently the only company in the UK that makes these. A 75 watt radiator should be powerful enough for a royal python vivarium.

Adult royal pythons feed mainly on mice. Two to three mice should be offered with tongs once a week. Remember, as with all snakes, to properly defrost the food items first. Royal pythons *can* on occasion have problems with feeding. Usually it is the young that exhibit this. Warming food items and emulating the movements of live prey may illicit a better response. If you are a beginner who is having trouble getting your snake to feed then ask the advice of an experienced keeper. It may be that they need to be force-fed with a tube or a 'pinkie press' (a nifty device for pumping minced up pinkies directly into a snake's stomach). A beginner should not attempt to force-feed a snake without the help of someone more experienced.

Any of the above snakes can make an excellent pet. The royal python is probably the most interesting, but for the beginner my recommendation goes to one of the hardy, attractive, and good-natured hognosed snakes.

Keeping Free Range *Nephila sp* in Captivity.

by Ray Gabriel

At some invertebrate livestock shows, various species of *Nephila* are often available from various parts of the world; Asian and New World species are less often found, but African species of *Nephila* sp are often seen for sale under the name *Nephila senegalensis*. Whether this is the correct identification I do not know, but for the purposes of this article I will refer to these as *Nephila sp.* I would like to apologise for the quality of some of the pictures, but these were taken pre-digital days, and in many instances I only had one to get some pictures.

Choosing, Housing and Acclimation;

Most of the African specimens seen for sale are from Tanzania, and are wild caught: as with most wild caught specimens, they are badly in need of re-hydration before they will start to spin large amounts of web. They will also need some space to do this in. If the spiders don't make a sufficient amount of web in a short period of time they tend to die.

If you decide to go for a *Nephila* sp the first thing you must decide is whether you are going to try to establish the spider in a cage, or in a very large bamboo cane hoop hanging from the ceiling, or let it free roam in a room to establish their webs where they want. I have tried many times to establish *Nephila* in cages and pet palls of all sizes, once in a hoop (well large triangle) but the best results were found by setting them loose in a room. As these spiders can spin very large webs I feel trying to keep them confined is half the problem when trying to acclimate them so I will concentrate on relaying my experiences with "free range" *Nephila*. I must also stress here that sometimes the *Nephila* will be found close to the floor in the mornings prior to making a web, if you are going to opt for free range *Nephila* ensure the room you are using is cat (or dog) free, most cats will try to play with the *Nephila* and either kill them or stress them so they make little web, which eventually leads to death.

I have always bought my *Nephila* specimens at the AES (Amateur Entomological Society) Exhibition at

Nephila sp. female

Nephila sp. male

Kempton Park Racecourse in late September early October each year. One of the reasons for doing this is that the first 24 hours of care are almost essential for establishing a *Nephila* in captivity and having Sunday free is a great help. Another reason for buying at this time of year is that *Nephila* specimens are almost always mated, and will produce egg-sacs once established. When first choosing your *Nephila*, look for a specimen with the largest abdomen. Try not to feel pity and select the weakest ones with the most shrivelled abdomens; these are most of the time (from personal experience) beyond saving. You should also try to find one where the back legs are complete, and none are missing. If you can get the spider to feed (ask the dealer if you can offer the spider a cricket to see if it will eat), if there are already uneaten food items in the spiders container when you are looking to buy, do not buy these, it shows they have lost most of the "fight" in them and most likely wont survive (unless they have been fed that day and the spiders have not had time to find and eat them). I have yet to see a mature male *Nephila* for sale, so as long as you are buying a specimen of more than 20mm leg diameter you will 99% of the time be buying a mature female.

When you get your spider home, let the spider out so it can climb somewhere high, the top of some curtains is my favoured place, but also because initially *Nephila* seem to do better if let loose very close to a window. After a couple of minutes offer the spider some water (a pinch of salt and sugar mixed in the water may help replace some salts when re-hydrating) from a bottle dropper such as a small rodent water bottle. If a bottle is not available then use a slow spray; it's OK the curtains *will* dry, but you might not want parents or partners watching you do this. Then offer a live cricket on the end of a pair of long forceps. This should be taken almost straight away. Keep an eye on the spider, and after the first cricket has been eaten (if the spider drops the cricket offer it again) repeat the process with water then another cricket, until the spider starts to refuse the food or it becomes too late at night. After a couple of hours, the *Nephila* should start putting down some web on the curtain/place it has been released. The following morning offer water again; a good sign of recovery is if the spider has moved, and put down more web. Anything from 4 days to 2 weeks later, a fine golden radial web will be found with a hungry *Nephila* sitting off centre. Once this web appears then keeping the spider is very straight forward. Don't try to cheat here and try to make things easy by offering large prey like locusts; the larger the prey the more chance of stressing the *Nephila* or the *Nephila* dropping the food.

I should note here that some people have had success establishing *Nephila* in large hoops suspended from the ceiling. If trying this the idea is to place the *Nephila* at the top of the hoop where it is suspended from the ceiling and feed and water her there, in the hope she will not wander, but make her web within the confines of the hoop. Personally, I have had no success establishing *Nephila* in this way, but I do know of people who have had successes doing this.

Feeding and Rearing

Larger *Nephila* will take adult black crickets as food offered at the end of a pair of long forceps; flies and moths caught in the web will also be eaten. By placing one or two crickets on the web and waiting for the female to find them is a good way to see how the female is getting on, after a while anything touching the web will be classed as "fair game". After eating, the females dispose of the remains by placing the old food pellets in an area at the edge of the web. A golden rule (pardon the pun) in the early stages of establishing a *Nephila* is that food equals web, the more food the spider will eat the more web will be produced. Once the *Nephila* is established then larger food items can be offered.

Nephila female in moult
- old prey items at the edge of web

Once the *Nephila* is established in a web, feed her very well and every couple of days spray water onto her and her web. *Nephila* seem to have evolved so that water in the form of rain (spray in captivity) runs down the abdomen onto the carapace and into the mouth (probably why they are most often seen in the wild head down). After a couple of weeks you might notice that the abdomen is swelling faster than the amount of food given and the amount of web produced, this is because there are eggs developing inside her. Most of the time I have noticed this swelling, an eggsac has been formed at the side of the web within a few days. The mother will feed almost straight away after the eggsac is completed, and continue to feed until the eggsac hatches. If well fed, a wild caught female can produce up to (from my experience) 4 eggsacs before she dies.

The *Nephila* spiderlings can then be housed individually in small clear ventilated containers and fed fruit flies or hatchling crickets, or left to disperse naturally in the room (house). The advantage of this later action is that they will find their own food and in the summer you will have a house devoid of flies and other winged insect visitors; the disadvantage is sticky golden webs in most corners of the rooms and windows.

Nephila spiderlings look quite different from the adults in both colour and shape, the legs do not appear so long and the front of the abdomen looks almost "hunchback" as it leans over the carapace. If well fed, spiderlings (from the first eggsac), can reach maturity (able to reproduce) within a year.

Moulting and mating

Moulting is a very interesting affair with *Nephila*. When close to a moult most *Nephila* stop feeding and suspend themselves by the tip of the abdomen to a thicker silk thread, attached to the edge of the web ensuring there is nothing underneath that she might come in contact with when they moult. The carapace splits in normal spider moulting fashion and the *Nephila* just seem to "fall out" of the old skin; they can hang there for almost 24 hours with all the legs hanging downwards while they recover and the new skin starts to harden up. Just like other spiders, this is a very strenuous time, so a bit of care is required for the few days after moulting. I do not attempt feeding until 7-8 days after the moult. Moulting is one of the most hazardous times when trying to keep *Nephila* in an enclosed container, the spider will drop from its skin during the moult and hit a branch or the side or base of the container and may be irreparably damaged if not killed.

Male *Nephila* are about 1/20th of the size of a female. The males of the African *Nephila* species are red and black, and are normally found on the outer edges off the female's web. Males rarely catch anything to eat for themselves, but scavenge from the uneaten remains of females or sometimes if they are very brave they will feed of the same freshly caught food item as the females are eating.

Radial web in living room-the spider can be seen slightly to the left of the tall plant in the corner of the room

Females of this species can reach roughly 12cm leg span from the hind legs to the front legs. The front 2 legs are black with a distinct yellow patch behind a distinct with a hairy "bottle brush" section on the tibia, the rear legs are yellow on the femur with the rest of the legs black and most of the hind tibia having the "bottle brush" hairs on the legs, The carapace is silver/white with a grey brown abdomen, with 4 pairs of whitish spots on the upper side; with white on the lower sides to underside.

One female (our first successfully established wild one we called the alpha female) that we kept free range in our living/dining room produced a web almost 2 metres in diameter . The guidelines for the web started at the ceiling light fitting in the centre of the dining room area, diagonally across to the wall onto the top of some book shelves; down onto the back of a dining room chair; onto the dining room table and back up to the ceiling. One guideline was placed across from the dining room table onto the opposite wall, but after we broke this guideline (to get past into the kitchen) the spider moved the guideline to the ceiling. If you have free roaming *Nephila* then this technique of damaging the guideline (one at a time) can get the web into a more suitable position. Younger immature females can be removed from their webs and placed in more manageable places (other windows for example) to make new webs.

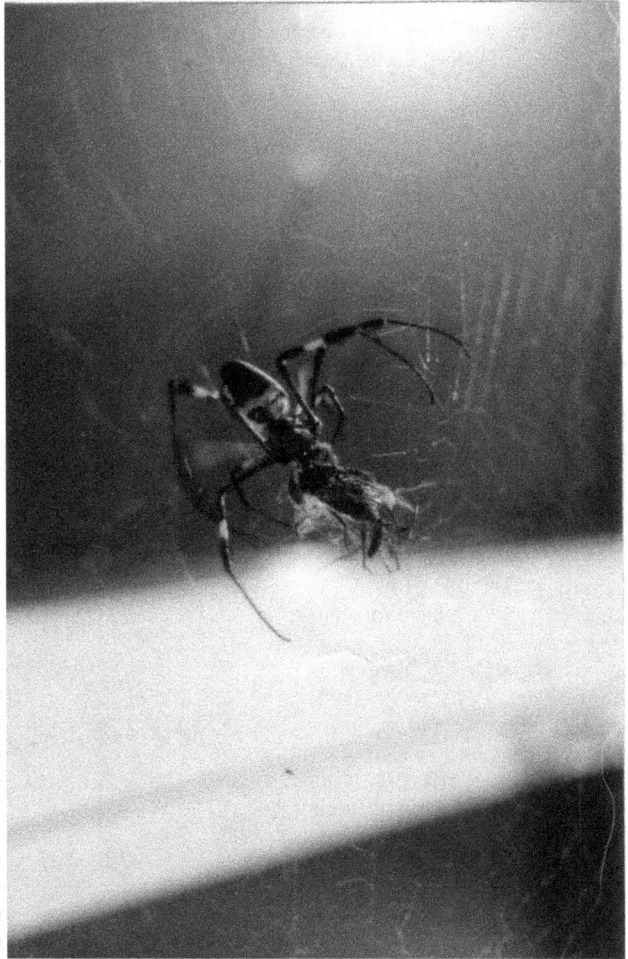

Nephila sp. feeding- with male feeding at t opposite end of the food item

In captivity males seem to "appear" overnight at the edges of females webs. Sometimes they can be found waiting at the edges of immature females' webs waiting for the females to moult. This form of mate guarding is not uncommon in arachnids, where the male waits for a female to either give birth then moult (some scorpions) or for the first maturing moult or a consecutive moult, so that he can be there to be the first to mate the female.

In the wild females of some *Nephila* species can be found very close together sometimes only a few inches apart (pers obs). A smaller female (we called her beta) from the same eggsac as the alpha female, made her web about 50cm away from the web of the alpha female. Both lived in this close proximity for around 4 months. A male was found on the alpha female's web but no male appeared on the beta female's. After 2 weeks or so when I was sure the alpha female had been mated I removed the male and placed her on the beta female's web late in the evening. When I came downstairs around 7:30 am the next morning I found

Nephila sp showing sexual diamorphism.

the (larger) alpha female on the beta female's web eating the beta female. I am not sure why this happened, but can only speculate that the alpha female "sensed" that the male had mated the beta female so ate the beta female to give her own offspring a better chance for survival. I am not sure if this occurs in the wild.

Eggsac Production

A few weeks after mating (again depending on the food supply), the female will produce an eggsac. The eggsac is a mass of externally loose golden thread, with a smaller more compact ball in the centre. Eggsacs hatch at room temperature in 3-4 weeks. Infertile eggsacs are hard in the centre of the silken mass.

I suggest spraying the eggsac once or twice a week. I think that dehydration in a warm room can also destroy the eggs in the centre of the eggsac.

Notes from the Wild

In Gambia I have seen Nephila sp webs about 3m from the ground spread between tree branches but most of the time I have found them from 1M to 2M from the ground.

In Panama most of the Nephila are found in shaded areas away from direct sunlight, while in Sri Lanka the largest Nephila specimen we observed was out in the open. While out on a night walk in Panama looking for Theraphosid spiders my partner spotted an orange Cupienus sp apparently hunting a Nephila on its web. Earlier that day we had spotted some commotion at the edge of a field, as we got closer we could see a hummingbird flying erratically around, as we got closer we found a destroyed Nephila web, the hummingbird was obviously collecting web to make a nest.

African Striped Grass Mice
by Oll Lewis

Africam striped grass mice is the name given to a number of different, but similar-looking, species of mice currently in the pet trade of the genus Lemniscomys. The two species most commonly kept as pets are *Lemniscomys striatus* the typical striped grass mouse and *Lemniscomys barbarus* the Barbary striped mouse. Typical striped grass mice are bigger as adults, with an average adult weight of 42.3g, than the Barbary striped mice at 26.2g. Both species of mice live for roughly the same time in captivity, just over 4 and a half years, but rarely survive for longer than one year in the wild.

Environment

African striped grass mice should be housed in groups. The mice are particularly sociable and in the wild live in very large colonies, so being housed singularly will stress them out. However, although striped mice do like to live together, if an enclosure is too small for the number of mice it houses the mice will compete with each other which, in extreme cases, can lead to fighting and deaths. There has to be adequate space for mice to get to areas of their enclosure away from other mice as well.

A covered glass tank 80cm by 40cm makes an adequately sized enclosure for a colony of less than 10 striped mice. The mice will appreciate several different floor levels as this will further increase the floor-space, provide space for the mice to get privacy, and give the mice more area for exercise. Space is very important when keeping striped mice because, in my experience, if you provide an exercise wheel the mice will usually ignore it, as they will not understand the concept. Neither do the mice appreciate being put in exercise balls, they will just sit in the ball looking vaguely confused. As well as the separate levels, provide several thick sticks to enable the mice to climb, and blocks of non-treated wood for the mice to chew.

The mice are intelligent, and are escape artists *par excellence* so if you use a wire cage make sure that the spaces between the bars are no bigger than 8mm and bear in mind that the mice can often work out the weak points in a cage and how to get out of it. One of these weak points are the doors in wire cages so it if using a wire cage it is a sensible precaution to twist some garden wire in a loop around both sides of the door so if they do manage to work out how the catch operates they will still not be able to open the cage doors.

There should be adequate depth in the tank to provide a substrate depth of at least 5cm. The substrate should be made from a thick layer of wood shavings covered with hay or a rodent safe paper or rag bedding. This enables the mice to dig and build communal nests which will keep them warm and insulated from drafts. It is very important to keep the mice warm, the species is used to ambient daytime temperatures of around 20 degrees Celsius in the wild so do not cope particularly well with lower temperatures and if the temperature is too low over a sustained period, the animals can die. Because of this the mice should always be kept indoors and in a heated, well insulated room, free from drafts.

Diet and health

The mouse's diet should be made of seeds grain with small amounts of fruit and animal protein. Seeds and grain should provide around 90% of the diet; this can be provided by a high quality commercial small rodent food that does not contain any additives or food colourings or if such a food is unavailable a substitute can be made from oats and sunflower seed in a 3:2 ratio.

The remaining 10% of the diet can be made from fresh fruit and animal protean in equal amounts. Certain fruits must be avoided because they can have a detrimental effect on the animal's health. Tomatoes, in particular, should never be fed to the mice because they contain oxalic acid, which can damage the mouse's small kidneys. Soft fruit, like bananas or strawberries should only be fed to the

animal as a treat and the bulk of fruits in the diet should be made up of harder fruits like apples and hard pears to help wear down the mouse's teeth. The proteins can be provided in the form of meal worms, shrimp pellets and on rare occasions small amounts of cooked lean meats like turkey. Do not over-feed animal proteins to the mice because too much fat will cause obesity.

Feeding bowls should be stainless steel or ceramic, and be wide and shallow enough to allow several mice to feed at the same time. For water, the mice prefer very shallow drinking bowls to bottles or deep bowls with steep sides, but will still take water from a bottle if no other alternative is provided. If using a bottle a glass bottle should be used, or the bottle should be guarded from the animals chewing by a mesh. Plastic bottles and bowls will be chewed by the animals, which can lead to digestive problems.

Food and water bowls have to be emptied and cleaned daily by either soaking in boiling hot water or use of a pet-safe disinfectant. If a disinfectant is used be sure to wash it out thoroughly before using the bowl for food. The reason why bowls have to be cleaned so regularly is because the mice will often excrete and urinate over their food. In the wild, this is to warn off mice from other colonies from coming near to their food, but in captivity this is redundant and a potential source of illness.

Socialising and Breeding

One of the main reasons given by pet shops as to why striped mice are not a more common pet is that they fight all the time. This is only the case when they are housed in tanks or cages far too small for the colony and have inadequate access to food and water. Pet shops and pet owners that are unaware of this, and treat the animals in exactly the same way they may treat any old mouse, have been known to be greeted by the sight of several dead mice, or mice with their tails chewed down to a stump. This is not normal behaviour

for the animals, and it will not be caused by a male fighting to gain a harem of females all to himself. It will be caused by one or two mice in the colony, either male or female, fighting over the short supply of resources. If you notice any gnawing of tails occurring, act swiftly and add more food and water bowls and move the animals to either a larger enclosure or add an extra floor to an existing enclosure, provided the clearance is not already too crowded.

If none of these options work you will have to split your colony up into separate enclosures. Before you do this keep a good watch over the animals for several days to observe if any natural groupings have occurred between the mice. If such groups have formed, it will be much less stressful for the mice if you move a group in its entirety rather than just moving random individuals.

Another population concern is breeding. If you provide African striped grass mice with the conditions outlined in this care-sheet as long as you have at least one male and female mouse they will breed. Breeding only takes place from mid-spring to near the end of summer but litters can be large and frequent. Typical grass mice will produce a litter of three to six young after a 1 month gestation period and can have as many as 3 litters in a breeding season. In practice, however, each female will only raise 1 litter each breeding season unless her babies die.

For the smaller Barbary striped mice gestation takes only 21 days and although they also produce a litter of three to six young they more frequently give birth to larger litters than the typical striped mice. A keeper does not have to provide any additional care for the young mice, which are born with a covering of short, thin hair and open their eyes a day or two after birth, other than providing extra food for the mother and for the babies when they start to eat solid food.

Small additional food and water bowls can be provided near the nest to ensure that the mother does not need to leave her litter for long in the early days after giving birth, but because the small mice will start to explore their environment only a few days after birth this will soon become unnecessary. In the weeks after birth carefully monitor the colony for signs of overcrowding. If offspring have to be taken away from the group this can be done without risk around 6 weeks or more after birth.

If you are going to enable your striped mice to breed, make sure you have a spare suitable enclosure set up in case your colony gets too large as it is likely a colony will more than double in size in one breeding season. If you intend to give colonies of mice to friends or acquaintances make sure that they are aware of the special needs of striped mice and inspect their proposed enclosures before-hand to make sure that they are suitable.

In Conclusion

African striped mice are good pets for people looking for something more interesting than a normal mouse. They are very intelligent mice, that provided you can meet their requirements, can thrive in captivity.

Insect Diets and the Role of Toxins in Food Plants

by Curtis Lakin

From time to time queries are raised on web forums regarding the toxicity of plants for feeding to herbivorous insects such as phasmids (stick insects) and to a lesser extent Orthoptera (grasshoppers and crickets).

Consideration of this topic should start with an understanding that each phasmid species has evolved over the millennia in its particular habitat in coexistence with its predators. In this evolution process species are likely to have developed to bio-accumulate (incorporate in the bodies cells) natural chemical toxins to various degrees to protect them against predators. As a consequence some species may be more tolerant to, or even crave for, particular toxins since it is part of the natural metabolic process of synthesising defence chemicals from these toxins. Whilst most caterpillars of Lepidoptera (butterflies and moths) seem to prefer one food-plant species to complete the metamorphosis life cycle, the same does not appear to be true of many species of stick insects and grasshoppers which seem to appreciate a mixed diet. Consequently such insects are often referred to as polyphagous in their dietary habits. Many species of Orthoptera such as the tettigoniids or bush crickets are also opportunistic omnivores and will also include in their diet live, freshly dead or dying insects and derive certain compounds to assist in the synthesis of chemical defences including the production of defence sprays.

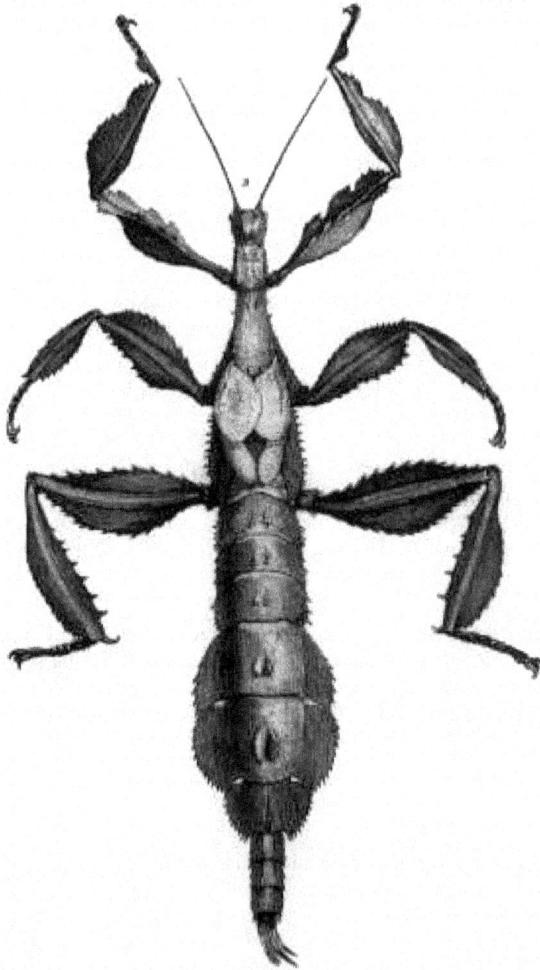

It should be borne in mind that the ability to metabolise the compounds from a particular plant will depend on the level of enzymes and related cofactors in the insects body at any time together with the rate at which it can synthesise these quickly. A practical example of this is that if you have been keeping a species like the Macleays Spectre or giant prickly stick (*Extatosoma tiaratum*) for some time on mature bramble leaves and then add fresh young shoots gradually this is less likely to be a stressor than if you switch from say a regular diet of oak to using bramble with only young shoots. The sudden switch may overload the animals' body without the means to metabolise the toxins. Another example is for a species which likes ivy, but you have been keeping it on bramble for some time. In this case my advice would be not to switch completely over to ivy in one go but introduce the ivy gradually. Whilst on the subject of ivy, I have heard people say that their insects are addicted to ivy – I would suggest that rather than addiction this is an inbuilt craving for toxins which are lacking through the inability to bio-accumulate these compounds from previous food offerings. Ultimately, if the insect has regular access to a food plant containing desirable toxins and reaches a satiated level of bioaccumulation it may then prefer to eat other things to maintain a balanced diet and gain nutritional factors that it cannot get from say ivy or privet.

The advantages of bio-accumulating toxins can serve both the individuals themselves and the population by making the insects more distasteful to predators. Females and males of a species may benefit directly in being more likely to be rejected by predators, but also the eggs from the resulting progeny may contain toxins (and may even be inoculated with the necessary enzymes) so that the next generation gains a good start. Research in some insects has supported this hypothesis by indicating that females can preferentially select males based on being able to detect characteristics associated with increased levels of toxins contained in their bodies.

In summary, the practical advice to those of us keeping Phasmids is that if you can offer a mixed diet of likely food plants each time you change the food that's probably the best policy both for the well being of the individuals and for the longer term health of the colony. If the insects refuse to eat some of the plants that's fine. It doesn't mean that at another time in its life it won't eat the plant. Very few species need very specific host food plants and a mixed diet or at least one alternate surrogate plant can only be beneficial.

A VIEW TO A KILLIE

On December the 7th, Jon Downes, Matthew Osborne and I travelled up to the Midlands to go to an auction. Not just any old auction; this one had fish. Nearly 500 lots of them to be specific.

TA Aquaculture UK holds two charity auctions every year, and does a sterling job raising money for local charities, whilst allowing hobbyists to meet up and chat about all things fishy, as well as the opportunity to get hold of some incredibly rare fish for stupid prices. Most of the lots were killifish, but there was something for everyone. Snakeheads, South American cichlids, tetras, livebearers, catfish from various families and anabantoids all made an appearance. Most were in group lots, either as sexed pairs or as groups.

We did arrive one hour late due to cock-ups on everyone's side, but this made no real difference to the days enjoyment. £500 exactly was raised by the event from door admissions, fish sales, the odds table, stand fees, donations and finally the raffle. The bar was superb with some local ales and great food to keep us going in the freezing cold conditions. Various stalls were arranged around the outside selling show tanks, plants, live foods and equipment. The British Killifish Association had a stand too, and it seemed like most of the people in the room were members, although I'm sure that their computerised presentation in the corner helped them recruit a few more.

Neither Jon nor I are much into killifish, but he was there for rare livebearers, and I was there for anything odd or rare. I have to admit that there was little there for me; many of the fish were captive bred by people present at the auction, so as most of the fish I am interested in are rarely bred. Something to remember for next time for sure, as is to bring a book if you are as rusty on rare livebearers as we were! Jon however had a smashing day as he came home with many pairs of *Heterandria formosa* as well as pairs of *Xiphophorus multilineatus* and *Xenotoca eiseni* for very little money.

The prices for some of the fish were just daft, the most expensive things were a pair of adult *Rivulus igneus* (a massive species) which went for £27, but this was far above the norm. Groups of various *Betta* species went for between £3-9 (including the stunning *B. imbellis* for £3); *Characodon audax*, known from one single spring in the wild for £5; 3 different undiscribed species of *Apistogramma* for less than £5 and *Zoogoneticus tequila* who are probably extinct in the wild for £3! You would pay easily £15 a pair for the *Apistogramma*, probably a lot more.

I would advise everyone who has even the slightest interest in the aquatic hobby to do to a local auction; they are always great fun and certainly get the heart racing as the bids fly!

The next charity auction is on 7th June. See you there. **MAX BLAKE**

book shelf

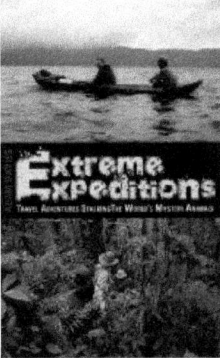

Extreme Expeditions: Travel and Adventures Stalking the World's Mystery Animals
Adam Davies

Paperback: 164 pages
Publisher: Anomalist Books LLC (7 April 2008)
Language English
ISBN-10: 1933665319
ISBN-13: 978-1933665313
Product Dimensions: 21.6 x 13.5 x 1.3 cm

There are too many armchair 'experts' in cryptozoology. People who never leave the comfort of their armchairs and pronounce to the world what can and can't exist without ever going to look for themselves. I recall a few years back when Jon Downes said that there couldn't possibly be a population of bigfoot type creatures living in the UK, he was met with an avalanche of vitriol from people who had never even been to England, berating him on daring to suggest that there were no bigfoots living just outside of Newcastle!

Then there are the *real* cryptozoologists. People who risk life and limb to hunt unknown animals at their own expense at the ends of the earth. Jon and I are of this rare breed and my good friend Adam Davis is another. Adam and I have know each other and followed each other's expeditions for years but it was only in 2008 that we finally went on an adventure together, hunting relic hominids in Russia.

Extreme Expeditions is Adams first book and chronicles his expeditions from 1998 up to 2007. Adam doesn't just talk the talk, he walks the walk; financing many expeditions by the sweat from his brow, and often going it alone into dangerous places. Here we follow him to the war-torn Congo in search of Mokele-Mbembe and braving mad, machete-wielding crowds. He treks through Sumatra after the Orang-Pendek regardless of earthquakes, and crosses the Gobi on the track of the Mongolian Deathworm. Perhaps the finest chapter is Adam's trip to Norway where, in Lake Seljord, he encounters a serpentine monster.

The book is not an in-depth examination of cryptids but an exciting and highly personal travelogue. My only criticism was that it wasn't long enough. I enjoyed having Adam along on our trip to Russia. He fitted into the team very well and it felt as if he had been on all my previous adventures. One memorable night in an abandoned farmhouse at 2.30 AM we may both have come within 12 feet of a wildman (almasty) as it stalked along the veranda outside the house.

We both ran out together, but whatever it was had gone. I found him to be courageous, resourceful, great at befriending locals, and having an excellent sense of humour. I was thrilled to hear he had written a book about his previous expeditions, and am overjoyed to say that the book did not disappoint. RICHARD FREEMAN

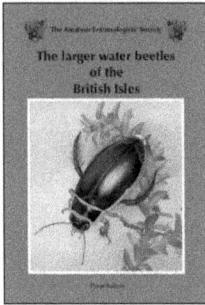

The Larger Water Beetles of the British Isles Peter Sutton
The Amateur Entomologist's Society ISBN 0-900054-74-3
Peter Sutton
80 pages,
b&w photos, line drawings.
Amateur Entomologists' Society

Before the Neo Nazis of political correctness banned children from pond dipping, it was one of the greatest childhood pleasures. I recall, as a boy in the 1970s many more people seemed to have garden ponds than they do today. It seemed like most children would be out on a Sunday afternoon with a net and bucket seeing what they could catch. I regularly took home frogspawn and was overjoyed when it transformed into tadpoles and then into frogs. This never did any damage to the wild populations, as we never took too much. There were always swarms of life left in the ponds and by taking some home top our own ponds we spread the population. One of the best finds a kid could ask for a great diving beetle or its larva. These were the tigers of the pond, top predators surpassed only by the grass snake. This little book came as quite a surprise to me as I didn't know just how many types of water beetle there were in the UK. The book covers no less than ten species that have wonderful names like 'enigma' and 'wasp'. I always thought the great diving beetle was the biggest water beetle in Britain but it turns out it is in fact the King Water Beetle.

There are excellent illustrations in both full colour and black and white as well as maps showing pre and post 1980 distribution for each species. I hope books like this will rekindle an interest in natural history that, in modern youth, seems lost in a putrid sea of video games, reality TV, and mediocre music. RF

Frogs and Toads of the South East
Mike Dorcas and Whit Gibbons
Paperback: 264 pages
Publisher: University of Georgia Press (1 Jul 2008) **Language** English
ISBN-10: 0820329223
ISBN-13: 978-0820329222

Charles Fort once said "*We shall pick up an existence by it's frogs.*" He was correct. These amphibians are some of the best indicators of a healthy environment. As barometers of the planet's well being they are not sending out good messages. Amphibians are dying out faster than any other group of animals.

Any book that encourages an interest in frogs and toads is fine by me, and this is a very good example. *'Frogs and Toads of the South East'* is a volume that bridges the gap between academic and popular. It is extensively illustrated with excellent colour photographs of each species, together with maps showing distribution, and in depth accounts of the biology of each animal. There are also nice touches like 'Did You Know?' facts interspersed throughout. 'Frogs and Toads of the South East' is as much a work of art as it is an important scientific reference book. What is striking is just how small the distribution of some of the species are. The dusky gopher frog (*Rana sevosa)* for example is know only from a handful of marshes in Louisiana, one in Alabama, and one in Mississippi.

The sheer amount of frog and toad species in this one area of the world is impressive. It is amazing just how many there are. It brings home how important these often-overlooked animals must be. One wonders how many have vanished before we even knew about them. RICHARD FREEMAN

HELP

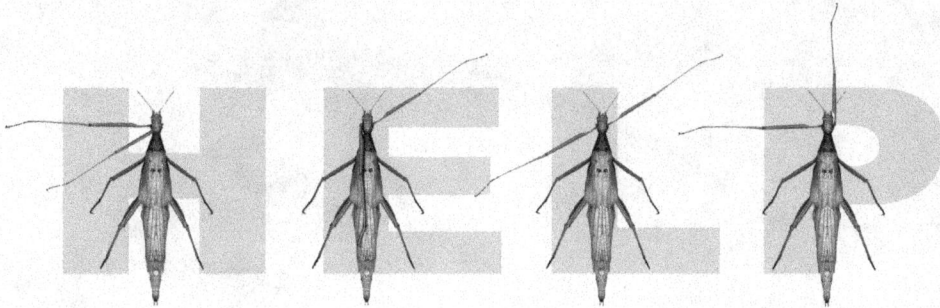

Dear Amateur Naturalist,

Is it true that the funnel-web spider's bite has no effect on cats and dogs but is deadly to primates? If so why?

Doreen Revel
Hinckley

Dear Doreen

Bizarrely this is true. Domestic mammals suffer little or no ill effects from the venom of a funnel-web spider. Conversely, humans and other primates like monkeys have a strong and often fatal reaction to it. This seems odd, as there are no naturally occurring primates in Australia.

Humans only arrived around 60,000 years ago. There *may* be a species of hominid known as the yowie living in Australia but these too would have arrived at a comparatively late date. The question arises as to why an Australian spider would have a venom that effects animals alongside which it did not evolve.

One idea is that the spider's venom is a very basic chemical compound that causes a non-specific reaction. It could be that the funnel-web spiders fill an ecological niche similar to tarantulas elsewhere. But were as tarantula venom is harmless to humans (and only effective against small prey) the funnel-web (that also hunts small prey in Australia) is - by fluke - deadly to primates.

All the best
Richard Freeman

Got a problem? Maybe we can help! Maybe we can't, but, if not, there is probably someone out there in the *Amateur Naturalist* readership who can.

Dear Amateur Naturalist,

I have been told that cane toads will eat dried dog food. I can't see any amphibian eating dried dog food myself. Is this on the level?

Adrian Stephens, Altherstone

Dear Adrian

I have also heard this odd story but never seen it happen in the flesh. I have kept many cane toads and, by way of experimentation, tried them on dry dog food and dry dog food that has been watered. Never once have they eaten it. Even when sprinkled with mealworms and crickets they will reject it spitting out any pieces they accidentally pick up. However I have seen film of feral cane toads been fed (in the manner one might feed pigeons in the park) on dried dog food in Northern Australia. Besides being horrified that this ecological-menace was being encouraged, I was amazed to see them gobble up the dried food like kids eating popcorn at a Saturday afternoon matinee.I fed mine on mealworms, locusts, and crickets dusted lightly with a multi-vitamin powder. In Australia though, they do have a reputation for eating anything.

Adios
Richard Freeman

Dear Amateur Naturalist,

My son Dan found a strange creature on the beach at Whitby last summer. I didn't have a camera at the time so I couldn't take a photo. I will try to describe it however; it was quite unlike any other animal I have ever seen. The creature was about six inches long and shaped like a slug. It was a dull brown in colour but I was covered in what looked like fur. The fur had an iridescent sheen like the rainbow effect of oil on water. It was hard to tell what end was the head.

Yours Tanya Kemp
Bedworth

Dear Tanya

What you found was a 'sea mouse'. The sea mouse is not a real mouse of course but a marine polychaete, a type of worm belive it or not! The 'fur'that gives it it's name is infact *chaetae,* chitinous bristles made from the same substanvce as invertebrate exoskeletons.
Cheers,

Richard Freeman
Dear Exotic Pets

The other night someone told me that sharks have to swim constantly and that they drown if they stop. Surely this can't be true?

Yours
Timothy Hands
Coventry

Well, yes and no. Some kinds of shark such as the great white, mako, oceanic whitetip and blue shark are what we call ram ventilators. They are generally pelagic (living in the open ocean) and need to keep swimming in order to pass water through the gills. These kinds of shark 'sleep' by shutting off part of the brain but keeping on swimming. However, other kinds of shark such as reef sharks and nurse sharks can rest on the sea bed and suck water through their gills.

All yours,
Richard Freeman

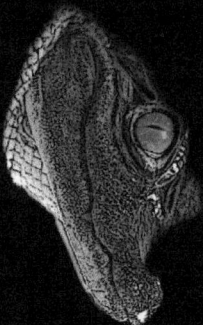

Jungle Jars

A complete invertebrate pet keeper's kit

http://www.metamorphosis.gb.com/

The Kit
- Jar
- Substrate
- Background picture
- Stick
- Plastic plant
- Vented lid
- Care sheet

Suitable for
- Mantids
- Beetles
- Cockroaches
- Assassin Bugs
- Grass hoppers
- Arboreal Spiders

Complete Kit for the Pet Bug of your choice!

THE AQUARIUM GAZETTE

The Aquarium Gazette
16 Potter Hill, Pickering, N.Yorks. Y018 8AA
Telephone 01751 472715
Email aquariumgazette@yahoo.com

Edited by Exotic Pets aquarium fish contributor David Marshall The Aquarium Gazette is the U.K.'s first bi-monthly aquarium magazine, available in both Word 2000 and PDF format, which can be purchased either on disc or for sending directly (PDF only) to your computer inbox.

Each Issue is comprised of 10 varied aquatic feature articles, written by well-known Aquarists', plus news and views from Aquatic Societies.

Issue 5 has articles that include African rasbora, the first U.K. spawning of *Acanthocobitis* loaches, breeding *Channa orientalis*, Water lilies, the threats from feral Snakeheads and a fascinating look at Australian crayfish.

Issue 6 has articles that include All about Koi, the first U.K. spawning of *Botia kubotai*, breeding Adoketa cichlids, an introduction to Bettas, the Doradids, Sea serpents and a 'cheeky look' at the world of fishkeeping.

Disc versions of these Issues are now available at the introductory price of £3.80 per Issue, inclusive of postage and packaging, or at £7.10 for the two Issue package. These Issues can be e-mailed directly to your computer inbox for the cost of £3 per Issue.

Payment can be made by sending a cheque (made payable to The Aquarium Gazette) to the above address or through our Paypal account (aquariumgazette@yahoo.com). If your preferred method of payment is through Paypal then please add 25p for a single Issue of the disc format or 50p for the two Issue disc package.

THE CENTRE FOR FORTEAN ZOOLOGY

So, what is the Centre for Fortean Zoology?

We are a non profit-making organisation founded in 1992 with the aim of being a clearing house for information, and coordinating research into mystery animals around the world. We also study out of place animals, rare and aberrant animal behaviour, and Zooform Phenomena; little-understood "things" that appear to be animals, but which are in fact nothing of the sort, and not even alive (at least in the way we understand the term).

Why should I join the Centre for Fortean Zoology?

Not only are we the biggest organisation of our type in the world, but - or so we like to think - we are the best. We are certainly the only truly global Cryptozoological research organisation, and we carry out our investigations using a strictly scientific set of guidelines. We are expanding all the time and looking to recruit new members to help us in our research into mysterious animals and strange creatures across the globe. Why should you join us? Because, if you are genuinely interested in trying to solve the last great mysteries of Mother Nature, there is nobody better than us with whom to do it.

What do I get if I join the Centre for Fortean Zoology?

For £12 a year, you get a four-issue subscription to our journal *Animals & Men*. Each issue contains 60 pages packed with news, articles, letters, research papers, field reports, and even a gossip column! The magazine is A5 in format with a full colour cover. You also have access to one of the world's largest collections of resource material dealing with cryptozoology and allied disciplines, and people from the CFZ membership regularly take part in fieldwork and expeditions around the world.

How is the Centre for Fortean Zoology organized?

The CFZ is managed by a three-man board of trustees, with a non-profit making trust registered with HM Government Stamp Office. The board of trustees is supported by a Permanent Directorate of full and part-time staff, and advised by a Consultancy Board of specialists - many of whom who are world-renowned experts in their particular field. We have regional representatives across the UK, the USA, and many other parts of the world, and are affiliated with other organisations whose aims and protocols mirror our own.

I am new to the subject, and although I am interested I have little practical knowledge. I don't want to feel out of my depth. What should I do?

Don't worry. We were *all* beginners once. You'll find that the people at the CFZ are friendly and approachable. We have a thriving forum on the website which is the hub of an ever-growing electronic community. You will soon find your feet. Many members of the CFZ Permanent Directorate started off as ordinary members, and now work full-time chasing monsters around the world.

I have an idea for a project which isn't on your website. What do I do?

Write to us, e-mail us, or telephone us. The list of future projects on the website is not   exhaustive. If you have a good idea for an investigation, please tell us. We may well be able to help.

How do I go on an expedition?

We are always looking for volunteers to join us. If you see a project that interests you, do not hesitate to get in touch with us. Under certain circumstances we can help provide funding for your trip. If you look on the future projects section of the website, you can see some of the projects that we have pencilled in for the next few years.

In 2003 and 2004 we sent three-man expeditions to Sumatra looking for Orang-Pendek - a semi-legendary bipedal ape. The same three went to Mongolia in 2005. All three members started off merely subscribers to the CFZ magazine.

Next time it could be you!

Project Kerinci, Sumatra - 2003
In search of the bipedal ape Orang Pendek

How is the Centre for Fortean Zoology funded?

We have no magic sources of income. All our funds come from donations, membership fees, works that we do for TV, radio or magazines, and sales of our publications and merchandise. We are always looking for corporate sponsorship, and other sources of revenue. If you have any ideas for fund-raising please let us know. However, unlike other cryptozoological organisations in the past, we do not live in an intellectual ivory tower. We are not afraid to get our hands dirty, and furthermore we are not one of those organisations where the membership have to raise money so that a privileged few can go on expensive foreign trips. Our research teams both in the UK and abroad, consist of a mixture of experienced and inexperienced personnel. We are truly a community, and work on the premise that the benefits of CFZ membership are open to all.

What do you do with the data you gather from your investigations and   expeditions?

Reports of our investigations are published on our website as soon as they are available. Preliminary reports are posted within days of the project finishing.

Each year we publish a 200 page yearbook containing research papers and expedition reports too long to be printed in the journal. We freely circulate our information to anybody who asks for it.

No. Each year since 2000 we have held our annual convention - the *Weird Weekend* - in Exeter. It is three days of lectures, workshops, and excursions. But most importantly it is a chance for members of the CFZ to meet each other, and to talk with the members of the permanent directorate in a relaxed and informal setting and preferably with a pint of beer in one hand. Since 2006 - the *Weird Weekend* has been bigger and better and held in the idyllic rural location of Woolsery in North Devon. The 2009 event will be held over the weekend 14-16 August.

Since relocating to North Devon in 2005 we have become ever more closely involved with other community organisations, and we hope that this trend will continue. We also work closely with Police Forces across the UK as consultants for animal mutilation cases, and we intend to forge closer links with the coastguard and other community services. We want to work closely with those who regularly travel into the Bristol Channel, so that if the recent trend of exotic animal visitors to our coastal waters continues, we can be out there as soon as possible.

We are building a Visitor's Centre in rural North Devon. This will not be open to the general public, but will provide a museum, a library and an educational resource for our members (currently over 400) across the globe. We are also planning a youth organisation which will involve children and young people in our activities.

Apart from having been the only Fortean Zoological organisation in the world to have consistently published material on all aspects of the subject for over a decade, we have achieved the following concrete results:

- *Disproved the myth relating to the headless so-called sea-serpent carcass of Durgan beach in Cornwall 1975*
- *Disproved the story of the 1988 puma skull of Lustleigh Cleave*
- *Carried out the only in-depth research ever into the mythos of the Cornish Owlman*
- *Made the first records of a tropical species of lamprey*
- *Made the first records of a luminous cave gnat larva in Thailand.*
- *Discovered a possible new species of British mammal - the beech marten.*
- *In 1994-6 carried out the first archival fortean zoological survey of Hong Kong.*
- *In the year 2000, CFZ theories where confirmed when an entirely new species of lizard was found resident in Britain.*
- *Identified the monster of Martin Mere in Lancashire as a giant wels catfish*
- *Expanded the known range of Armitage's skink in the Gambia by 80%*
- *Obtained photographic evidence of the remains of Europe's largest known pike*
- *Carried out the first ever in-depth study of the ninki-nanka*
- *Carried out the first attempt to breed Puerto Rican cave snails in captivity*
- *Were the first European explorers to visit the `lost valley` in Sumatra*
- *Published the first ever evidence for a new tribe of pygmies in Guyana*
- *Published the first evidence for a new species of caiman in Guyana*

Other books available from
CFZ PRESS

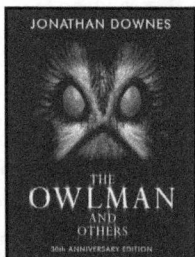

THE OWLMAN AND OTHERS - 30th Anniversary Edition
Jonathan Downes - ISBN 978-1-905723-02-7

£14.99

EASTER 1976 - Two young girls playing in the churchyard of Mawnan Old Church in southern Cornwall were frightened by what they described as a "nasty bird-man". A series of sightings that has continued to the present day. These grotesque and frightening episodes have fascinated researchers for three decades now, and one man has spent years collecting all the available evidence into a book. To mark the 30th anniversary of these sightings, Jonathan Downes has published a special edition of his book.

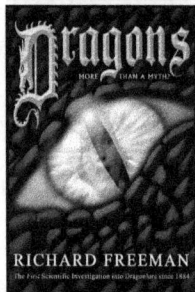

DRAGONS - More than a myth?
Richard Freeman - ISBN 0-9512872-9-X

£14.99

First scientific look at dragons since 1884. It looks at dragon legends worldwide, and examines modern sightings of dragon-like creatures, as well as some of the more esoteric theories surrounding dragonkind.

Dragons are discussed from a folkloric, historical and cryptozoological perspective, and Richard Freeman concludes that: "When your parents told you that dragons don't exist - they lied!"

MONSTER HUNTER
Jonathan Downes - ISBN 0-9512872-7-3

£14.99

Jonathan Downes' long-awaited autobiography, *Monster Hunter*...

Written with refreshing candour, it is the extraordinary story of an extraordinary life, in which the author crosses paths with wizards, rock stars, terrorists, and a bewildering array of mythical and not so mythical monsters, and still just about manages to emerge with his sanity intact.......

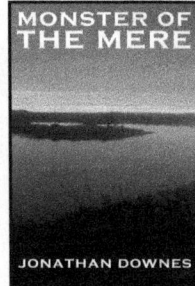

MONSTER OF THE MERE
Jonathan Downes - ISBN 0-9512872-2-2

£12.50

It all starts on Valentine's Day 2002 when a Lancashire newspaper announces that "Something" has been attacking swans at a nature reserve in Lancashire. Eyewitnesses have reported that a giant unknown creature has been dragging fully grown swans beneath the water at Martin Mere. An intrepid team from the Exeter based Centre for Fortean Zoology, led by the author, make two trips – each of a week – to the lake and its surrounding marshlands. During their investigations they uncover a thrilling and complex web of historical fact and fancy, quasi Fortean occurrences, strange animals and even human sacrifice.

**CFZ PRESS, MYRTLE COTTAGE,
WOOLFARDISWORTHY BIDEFORD,
NORTH DEVON, EX39 5QR
w w w . c f z . o r g . u k**

Other books available from
CFZ PRESS

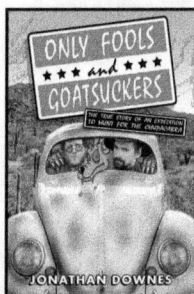

CFZ PRESS

ONLY FOOLS AND GOATSUCKERS
Jonathan Downes - ISBN 0-9512872-3-0

£12.50

In January and February 1998 Jonathan Downes and Graham Inglis of the Centre for Fortean Zoology spent three and a half weeks in Puerto Rico, Mexico and Florida, accompanied by a film crew from UK Channel 4 TV. Their aim was to make a documentary about the terrifying chupacabra - a vampiric creature that exists somewhere in the grey area between folklore and reality. This remarkable book tells the gripping, sometimes scary, and often hilariously funny story of how the boys from the CFZ did their best to subvert the medium of contemporary TV documentary making and actually do their job.

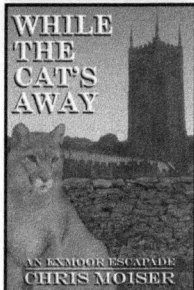

WHILE THE CAT'S AWAY
Chris Moiser - ISBN: 0-9512872-1-4

£7.99

Over the past thirty years or so there have been numerous sightings of large exotic cats, including black leopards, pumas and lynx, in the South West of England. Former Rhodesian soldier Sam McCall moved to North Devon and became a farmer and pub owner when Rhodesia became Zimbabwe in 1980. Over the years despite many of his pub regulars having seen the "Beast of Exmoor" Sam wasn't at all sure that it existed. Then a series of happenings made him change his mind. Chris Moiser—a zoologist—is well known for his research into the mystery cats of the westcountry. This is his first novel.

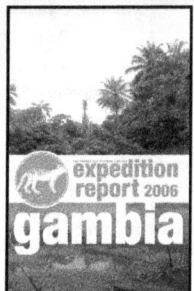

CFZ EXPEDITION REPORT 2006 - GAMBIA
ISBN 1905723032

£12.50

In July 2006, The J.T.Downes memorial Gambia Expedition - a six-person team - Chris Moiser, Richard Freeman, Chris Clarke, Oll Lewis, Lisa Dowley and Suzi Marsh went to the Gambia, West Africa. They went in search of a dragon-like creature, known to the natives as `Ninki Nanka`, which has terrorized the tiny African state for generations, and has reportedly killed people as recently as the 1990s. They also went to dig up part of a beach where an amateur naturalist claims to have buried the carcass of a mysterious fifteen foot sea monster named 'Gambo', and they sought to find the Armitage's Skink (*Chalcides armitagei*) - a tiny lizard first described in 1922 and only rediscovered in 1989. Here, for the first time, is their story.... With an forward by Dr. Karl Shuker and introduction by Jonathan Downes.

BIG CATS IN BRITAIN YEARBOOK 2006
Edited by Mark Fraser - ISBN 978-1905723-01-0

£10.00

Big cats are said to roam the British Isles and Ireland even now as you are sitting and reading this. People from all walks of life encounter these mysterious felines on a daily basis in every nook and cranny of these two countries. Most are jet-black, some are white, some are brown, in fact big cats of every description and colour are seen by some unsuspecting person while on his or her daily business. 'Big Cats in Britain' are the largest and most active group in the British Isles and Ireland This is their first book. It contains a run-down of every known big cat sighting in the UK during 2005, together with essays by various luminaries of the British big cat research community which place the phenomenon into scientific, cultural, and historical perspective.

**CFZ PRESS, MYRTLE COTTAGE,
WOOLSERY, BIDEFORD,
NORTH DEVON, EX39 5QR
www.cfz.org.uk**

Other books available from
CFZ PRESS

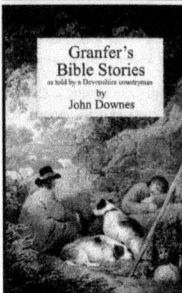

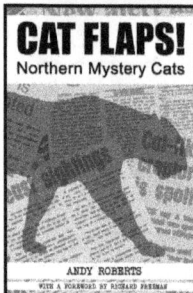

Other books available from
CFZ PRESS

Other books available from
CFZ PRESS

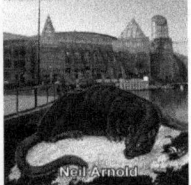